I0709742

Hide and Seek

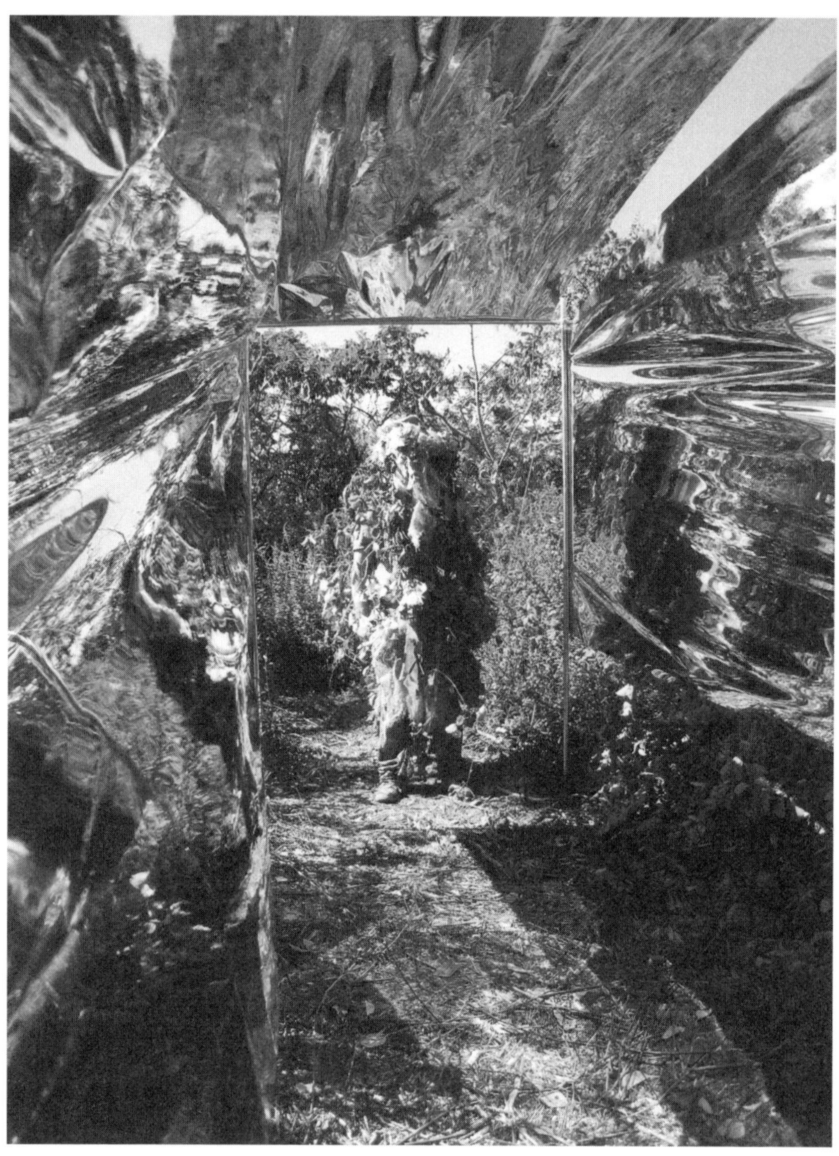

Hide and seek. Hanna Rose Shell, *Blind*, 2008. Courtesy of the artist.

Hide and Seek

Camouflage, Photography, and the Media of Reconnaissance

Hanna Rose Shell

ZONE BOOKS · NEW YORK

2012

Publication of this book has been aided by a grant
from the Millard Meiss Publication Fund of the
College Art Association.

MM

ZONE BOOKS
1226 Prospect Avenue
Brooklyn, NY 11218

Printed in the United States of America.

Distributed by The MIT Press,
Cambridge, Massachusetts, and London, England

Library of Congress Cataloging-in-Publication Data

Shell, Hanna Rose.
    Hide and seek: camouflage, photography, and the
media of reconnaissance / Hanna Rose Shell.
        p.   cm.
    Includes bibliographical references and index.
    ISBN 978-1-935408-22-2 (alk. paper)
    1. Photography — Special effects. 2. Photographic
reconnaissance systems. 3. Hidden camera photography.
4. Art and camouflage. I. Title.

TR148.S54 2012
770 — dc23
                                2011037201

Contents

Dedicated to Sophie S., Murray, and Sophie M.,
whose strength of character and rootedness in history
energize my every day and word.

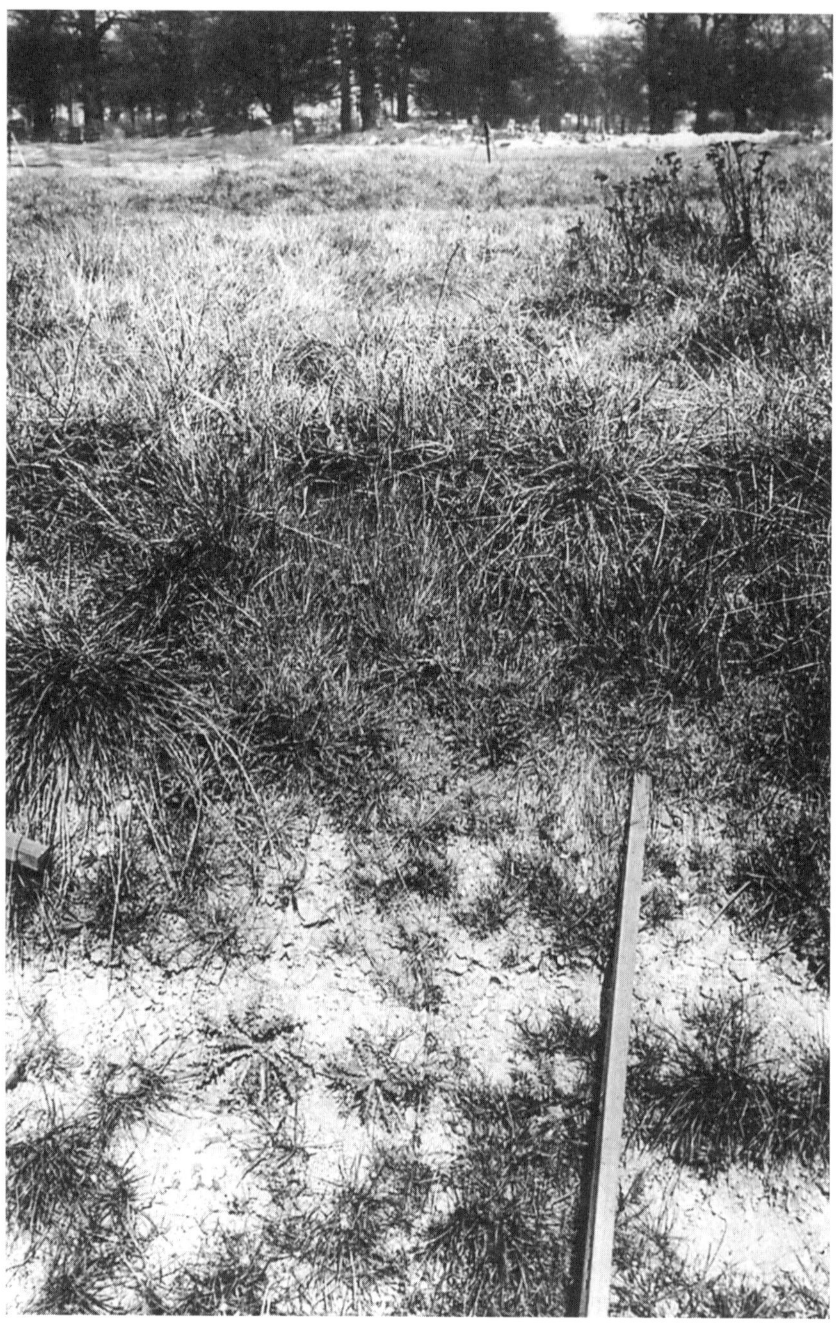

Figure P.1—Sniper in grass. Instructional photograph distributed by the British War Office, ca. 1916. Reprinted courtesy of the Imperial War Museum [Q17729].

Preface

A photograph captures everything in that it reveals nothing.

This pedagogical image (figure P.1) was first distributed by the British War Office in 1916 to members of the British Army and its allies. A wooden stick reaches into the photographic frame to direct our attention to what the image's title implies should be a craftily concealed human being: a model sniper, lurking in tall, weedy grass. The invisible subject of instruction is camouflage.

But where is the sniper? *Is* there a sniper? No matter where the viewer looks, she is hard-pressed to find traces of a person encoded in the emulsion. The silver nitrate grain expresses no tell-tale shadow, hue, or texture. Differentiating the foreground from the background, the human from the natural, is difficult, if not impossible. Even its scale is hard for some viewers to gauge. The photograph is uncomfortably hard to read, its hider, who is also a seeker, even harder to see. When the photograph is blown up big, some viewers profess to see the circular end of a gun dark against the grass. Others see several possible giveaways, while still others see only grass and a deceitful stick, pointing at nothing.[1]

If the sniper *is* there, rifle cocked, eyes peeled, he is observing the world unseen. Given that he cannot be found, this image would be photographic documentation of a job exceedingly well done. But perhaps there was never a man hiding in these weeds at all; we're staring at an empty field. Either way, there's something to learn, for both the World War I soldier in training—for whom the photograph was originally produced—and for the modern-day civilian. To study camouflage is to investigate how we look at the world and how we conceal ourselves in and among photographs of that world. The logic of camouflage is predicated on the assumption that *not showing up* is, at times and places, both a strategic necessity and a worthy aspiration.

Let's assume there *is* a man there somewhere. In producing this unsettling pedagogical instrument, the War Office aimed to suggest that if and when you need to hide, you should emulate the hidden human subject, the alleged "sniper in grass." Immerse yourself and your own sighting devices into the landscape so that you appear to disappear within its sculptural and textural contours, whatever those may be, wherever you are.

As it happens, the protophotographic environment depicted in this photograph is a scrubby patch of Kensington Gardens, London, which served as the Special Works School research facility during World War I. If you needed to get lost in a similarly gravelly, grassy habitat, you would garnish yourself with local foliage, add face paint, and burrow behind the grass, under the rocks. Disappearance is, within this pedagogical regime, always *from* something—from photographic exposure or from the perceptual capabilities of the photograph's eventual viewer. Disappearing is the effect of receding into the background of one's projected photographic habitat, that is, of not being visible in a surveillance photograph that might be made at this or a future moment. Do as the sniper does, and maybe you will fool the camera, too, and countless and unknown future others.

Get the better of a photograph? This is no easy task. Ever since William Henry Fox Talbot, inventor of the first photographic negative, characterized photography as "the pencil of nature" in 1844, critical observers from Oliver Wendell Holmes, to André Bazin, to Susan Sontag, to Roland Barthes have articulated how conditions of photographic production influence the photograph's status as a "trace" or "skin" of an original (or protofilmic) form. Philosopher and founder of semiotics Charles Sanders Peirce called attention to the peculiar twinned epistemological status of the photographic negative. The photograph, like the mounted skin of an animal, both materially derives from the live organism with which it is associated (a relationship described as "indexical"), and resembles that organism, possessing some of its qualities (a relationship described as "iconic"). It is because the relationship is simultaneously indexical and iconic that photographs are so widely used in evidentiary and illustrative purposes, as in criminology and detective work.

A straightforward understanding of a photograph as a direct "trace" of a protofilmic form guided the development of everything from medical atlases to criminological archives. Art historians and philosophers of science interrogated a related "truth to nature" photographic episteme.[2] Meanwhile, cultural critics argued that notions of "objectivity" and "indexicality" facilitated the development of multiple fields of science and technology in the late nineteenth and early twentieth centuries.[3] Such work effectively details how, despite repeated claims to the contrary, both the photographer herself and the larger social and economic forces guiding her work shape the final photographic image. Reception and consumption play a crucial role in determining whatever signifying function the image will have.

Back to the photograph in question: The presence of the stick leads us to assume that if we look closely enough at the spot to which it points, we will triumphantly discover our man. But before us there is no foreground or background; just ambiguous ground. Did the sniper not show up on the scene, as well as not showing up visually in the photograph? In that event, the brush, the face paint, the careful body positioning, all would have been wasted. Perhaps this is a pretense of a pretense, a double deconstruction of representation. Maybe we are the ones who have been tricked. Exposed, as it were.

What is the configuration of self, environment, and experience that leads one to hide inside a photograph? To conceal oneself? To become something else? To see everything? To not show up? These questions are conjoined motivations for, and expressions of, the logic of camouflage.

Camouflage Media

Here's one way of not showing up. In 1918, a young American artist and infantryman named Homer Saint-Gaudens invented a new machine in response to a pressing problem. A shortage of blankets had beset the American Expeditionary Forces installed along the Western Front, and had threatened the war effort. Until that hard winter, secondhand wool had been collected in British and American cities for recycling; at specialized factories known as "shoddy mills," this reclaimed material was transformed into new fabrics, primarily bolts of boiled wool dyed olive, brown, or gray-green. These blankets were widely distributed among the infantry and had provided both warmth and concealment, but by January 1918, raw materials and labor were scarce, and factory production stalled. It was cold, it was damp, and soldiers looked into the sky and saw planes equipped with cameras flying overhead.

Saint-Gaudens, son of the renowned sculptor Augustus Saint-Gaudens, had a reputation for making do in a pinch. Early in 1917, when fabric had likewise run out at Camouflage Training Camp in Plattsburgh, New York, he had dug up faded navy overcoats, remnants from the American Civil War, to clothe the new enlistees.[1] Soon thereafter, Saint-Gaudens sailed with the rest of the Fortieth Engineers across the Atlantic, landing at Brest, France. Upon arriving at headquarters in Dijon, the energetic thirty-seven-year-old proposed a solution to the looming blanket shortage crisis. Captain Saint-Gaudens's ingenius invention would turn old newsprint, discarded letters, war bulletins, and bound volumes whose spines had worn out into camouflage material. Run through the press-and-dye works, shredded paper would reemerge as thick sheets of woven material in environmentally coordinated shades of gray, green, and brown.

Effectively deployed, the camouflage material produced by Saint-Gaudens' book-devouring machine could make people—soldiers or civilians—disappear into photographs much like the photograph of the sniper hidden in the grass, only to be revealed (or not revealed) later. As *Scientific American* reported in the spring of 1918, the blankets of camouflage material were "tinted like the surrounding grass and used as a cover for the bodies of men going up to the front."[2] Infantrymen had been supplied a textile skin. In an almost preternatural process of replicating the landscape, the "material," as the troops called it, was grafted onto observation posts, artillery factories, trench parapets, and gun batteries. From printed paper grew a second nature.[3]

The term "camouflage" was absorbed into English and American lexicon almost immediately after its introduction. Coined in 1914 by French general and artist Lucien Victor Guirand de Scovela, entering into the English language early the following year, the term "camouflage" came into being as a way to refer to systematic dissimulation for the purposes of concealment from photographic detection. The etymology is twofold: On the one hand, "camouflage" harkens back to the nineteenth-century French word *camou-flet*, which refers either to a primitive land mine that creates potent underground explosions without surface rupture or to a tiny smoke bomb that explodes when placed into the nose of an unwitting victim. "Camouflage" is also related to the medieval Italian *camuffare*, meaning "to make up."[4]

The sort of systematic dissimulation that "camouflage" implies involves both individual and collective practices. It is effected through human mimicry of natural forms—mimicry, that is, in the sense of visual resemblance, rather than similitude in any ontological sense—as well as through the construction of decoy military forms. Of course, strategic concealment itself had been a component of warfare long before 1914; one need only consider Homer's history of the Trojan Horse or Birnam Wood's march to Dunsinane in *Macbeth*. In the former, the Greek army gains entry to Troy through use of a giant decoy horse, a move widely understood as treachery; in the latter, an invading army disguises its approach by concealing its human soldiers as trees from a nearby forest, treated as simply an effective strategy.[5] But not until the twentieth century were such practices explicitly set forth and institutionalized as a body of interwoven scientific theories and artisanal practices.

Blame the camera, perhaps. Camouflage was both a potent response to photography's practical and theoretical effects on biology, military technology, and the arts and an instigator of its further development. Trench warfare was made infinitely more complicated by the new possibilities of aerial photography. Artists and cultural critics, as well as politicians and military leaders, heralded photography's heightened significance. To combat this new and in a sense increased threat of being seen by the reconnaissance lens—which included not only aerial cameras, but also periscopes and sniper scopes—*camoufleurs* brought to bear craft skills from backgrounds as diverse as taxidermy, architecture, set design, and portraiture. And they did so against a cultural backdrop in which the ability of human understanding and technologies to expose heretofore misunderstood aspects of natural history was simultaneously celebrated and decried. Could the camera help prove the truth of natural selection? Or did humans, as others alleged, fool themselves into thinking the products of the lens could further any kind of usable knowledge?

Saint-Gaudens himself likened the practice of self-concealment via camouflage to a woman's toilette. Both entail acts of deception in anticipation of the possibility of a spectator's gaze. He recalled installing over his body and his peers "bunches of burlaps to produce a texture like their surroundings.... It would thin out at the side so as to blur the spot into the surroundings, as a girl blends rouge into her face."[6]

Scientists, soldiers, artists, designers, and, most likely, this book's readers have varied associations with this word "camouflage," many of which seem to have little explicit connection to photography.[7] Consider for a moment some of the referents to which "camouflage" is often applied: a decorator crab at the aquarium that dresses itself in the detritus of its underwater habitat; a military tank in action, displaying its variegated colors as it rumbles down a Baghdad street; a World War I howitzer on display at London's Imperial War Museum; or children's bathing trunks, machine pockmarked with splotches of green and gray, and on sale at the local mall for $4.99 (figure 1.1). Say "camouflage" to an octopus physiologist or a high-school freshman taking an introductory biology course, and you will hear about specific animals and their evolutionary histories: for example, the cuttlefish, the arctic hare, the leaf bug, or the hawk moth (figure 1.2).[8]

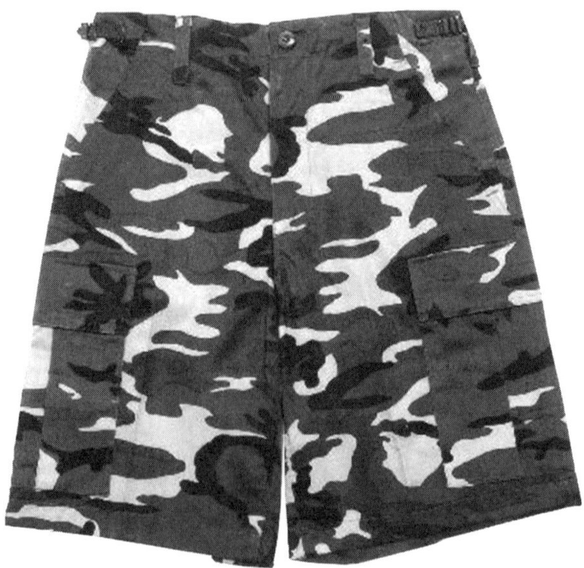

Figure I.1—Disruptive pattern material. The M81 Woodland Battle Dress Uniform (BDU) was first issued to American armed forces troops in 1981 and remained standard until 2005. Today, M81-patterned textiles are found in a range of civilian contexts, from shopping malls to hunting grounds. Photo by the author.

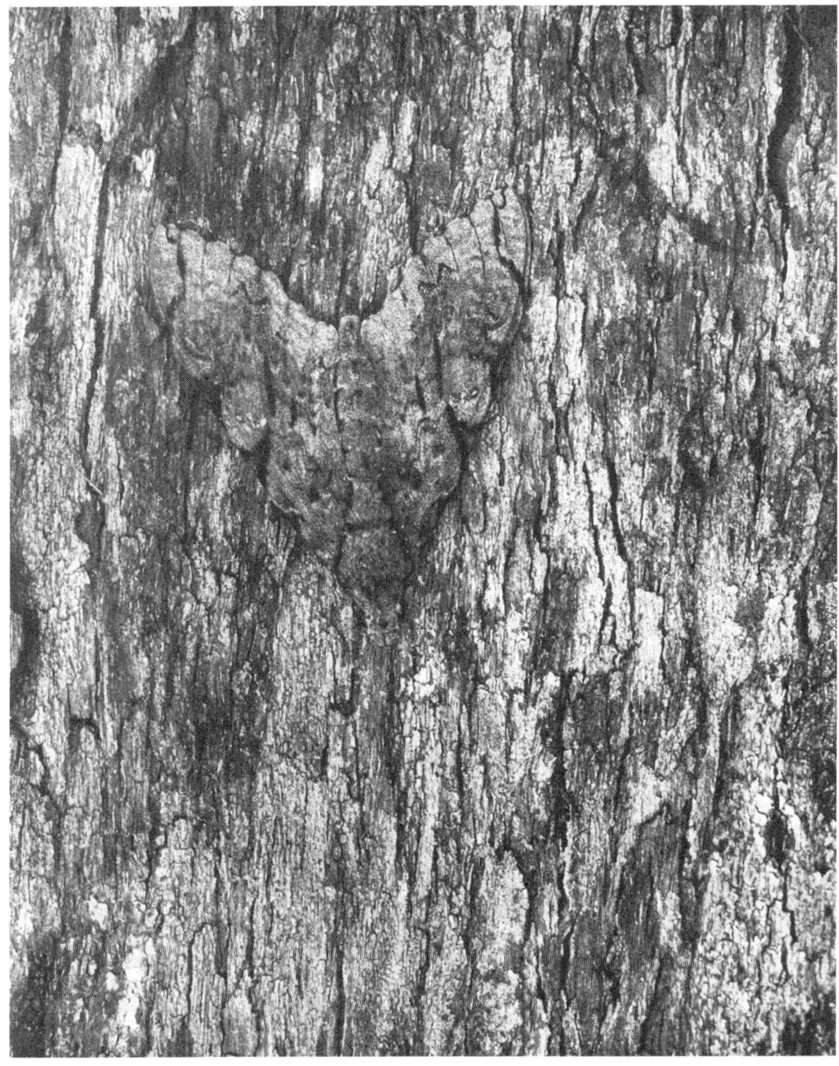

Figure I.2 — A moth in hiding. A hawk moth (*Xanthopan morgani*) blends into tree bark, although not well enough to conceal itself from the photographer's eye or the camera's lens. Reproduced from Hugh Cott, *Zoological Photography in Practice* (London: Fountain Press, 1956), p. 243.

The reference might also be not to a singular object or organism, but rather to a more general visual aspiration or a condition of psychology. You can drape yourself in trademarked camouflage patterns to articulate military fervor or more pacifist—even anarchist—inclinations. Generally speaking, an army or a paintball troupe wears camouflage because it offers some degree of environmental concealment in specific conditions. Yet at the same time, these identical uniforms clearly demarcate the individual as a member of a group.

There is irony there. Camouflage serves as a moniker both for blending in and for standing out. To be effective, the owner of a particular piece of camouflage must be readily identifiable, which is why, when Finnish authorities accused the Russians of stealing their trademarked M/05 camouflage pattern in 2008, no one could deny the potential for trouble, even as the charges were quickly brushed aside.[9] If Russian forces crossed their country's eight-hundred-mile border with Finland, it would be impossible to tell who was who. Contemporary artists have interrogated the apparently simultaneous construction and eradication of tropes of personal and national identity. In Andy Warhol's final series of self-(concealment)-portraits, produced in the mid-1980s, his face ghostly is impressed into a camouflage pattern rendered in Technicolor hues.[10] Thomas Hirschhorn erects cacophonous installations, assemblages of mountains of teetering camouflage bric-a-brac, military trappings, and consumer trinkets, assembled via eBay and affixed with duct tape into ironic embraces; fashion meets war in the realm of *Utopia, Utopia = One World, One War, One Army, One Dress.*[11]

According to an often-cited quotation from Gertrude Stein, Pablo Picasso—three years after proclaiming himself the inventor of the collage (or *papier collé*)—claimed that camouflage derived from cubism. "It was at night, we had heard of camouflage but had not yet seen it and Picasso, amazed, looked at it and then cried out, 'yes it is we who made it, that is cubism.'"[12] Before him, parading down a Paris street, surrounded by crowds, was a tank dressed in interlocking geometric forms of green, brown, and black.

Stein's gloss on Picasso has been cited by cultural, art, and even military historians as proof of the association between camouflage and cubism. Her assessment has received a remarkable amount of attention, some scholars using it to elaborate on the idea that camouflage, like cubism, is expressive of the modernist impulse in both art and politics.[13] But more thought-provoking still is the fact that the camouflaged tank, the "it" to which Picasso referred, was in fact perfectly visible—to Picasso and to Stein alike—standing out from, rather than blending into, the scene. What they saw was power on display, an expression of a state of war. Meanwhile, as the crowds watched "Picasso's" tank barrel through Paris and journalists commented on its unique painted patterns, there was a different kind of camouflage innovation in the making. And this camouflage innovation was effacive, rather than expressive in form, function, and genealogy (figure 1.3).

To soldiers themselves, those men and women long uniformed in the curvilinear pattern that sparked Picasso's boast, "camouflage" is a critical subject of study taught in basic training. In this sense, camouflage cannot be preprinted or packaged; rather, it is meaningful as a way of seeing, being, moving, and working in the world. It is a form of cultivated subjectivity. As such, it is an individuated form of self-awareness that is also part of a network of institutional practices.

Camouflage unfolds in time and space, across disciplinary and discursive boundaries, as an adaptive logic of escape from photographic representation. *Hide and Seek* investigates the structuring principles of visual pedagogies and material practices of strategic concealment between the first publication of Darwin's *The Origin of Species* and the close of World War II, and reveals the conditions under which camouflage media came to be widely produced and disseminated.

In doing so, I propose three historically and conceptually linked species or structural formations of camouflage. These I call the "static," the "serial," and the "dynamic," all terms that reflect distinctive types of photographic reconnaissance at stake in the logic of their generation. These species formations have developed in sequence, but have not replaced one another. The evolution of each

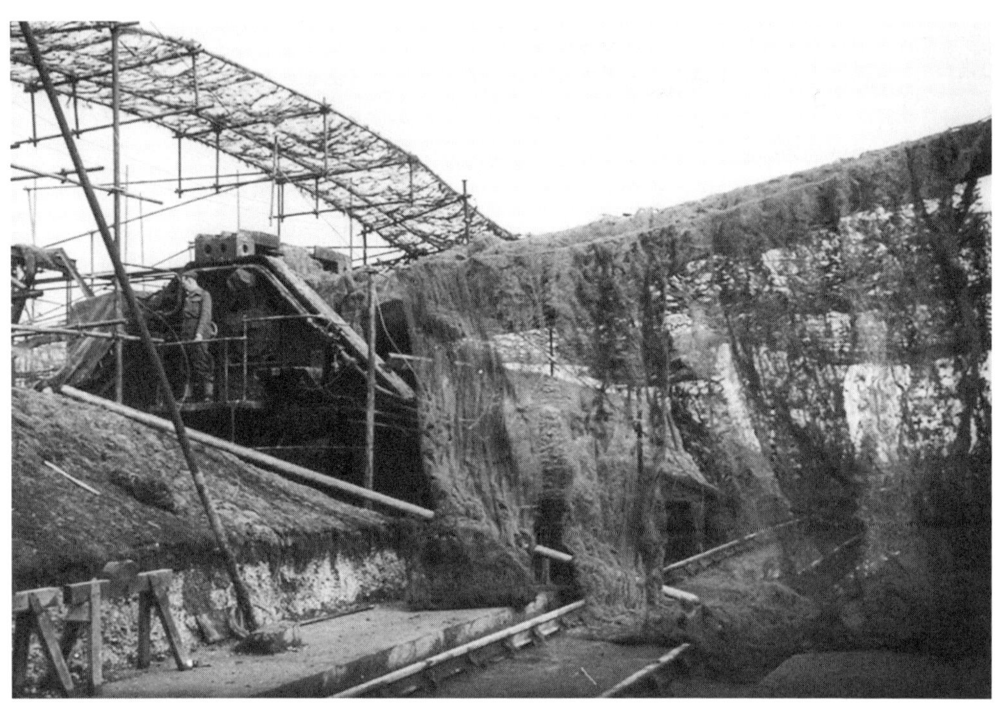

Figure I.3—**Enveloped by netting**. Railway sidings festooned with hanging vines of camouflage material obscure fourteen-inch railway guns from an aerial view at St. Margaret's-at-Cliffe, Kent. Reprinted courtesy of the Imperial War Museum [H7925].

new species has been dependent on the existence of a previous one, which has remained in operation both on its own terms and as constitutive of later forms. The three structural formations of camouflage have coexisted since the mid-twentieth century, and yet each contains aspects of the previous methods of photographic camouflage. As such, each is an individuated form of self-awareness that is also part of a network of institutional practices.

Camouflage escapes representation by design, so it is unsurprising that the traces of its unfolding over time are often elusive. Meanwhile, the forms and media of camouflage keep changing as environments and surveillance technologies evolve. This book traces the history of camouflage from the material forms it has left behind—photomontages, paintings, paper blankets, stuffed rabbits, ghillie suits, and instructional films.

Chapter 1, "Productive Mimesis and the Art of Disappearance," examines static camouflage, strategic concealment in relation to the instantaneous photograph (as in figure P.1), as it emerged in the domains of natural history and figurative art. It focuses on the work of a single person, Abbott Thayer, whose evolving ideas and media practices provide a means of understanding camouflage in the absence of change or the passage of time. Chapter 2, "Mending the Net," considers the emergence of serial camouflage as it took shape in the form of the "netting" that began to become synonymous with modern warfare, starting in World War I. Serial camouflage emerged in a military context as an overarching set of practices forged by a shifting collective of human and nonhuman actors united in the aim of circumventing aerial reconnaissance technologies. Chapter 3, "How Not to Be Seen," examines dynamic camouflage, strategic concealment effected in response to real-time filmic surveillance in which camera positioning and movement are unknown. This chapter posits the sniper as a model for land warfare and cinema spectatorship by the mid-twentieth century. Chapter 4, "Subject to Change," examines the convergence of static, serial, and dynamic modes of camouflage media in the service of what might be called the chameleonic impulse, that is, the human drive to represent,

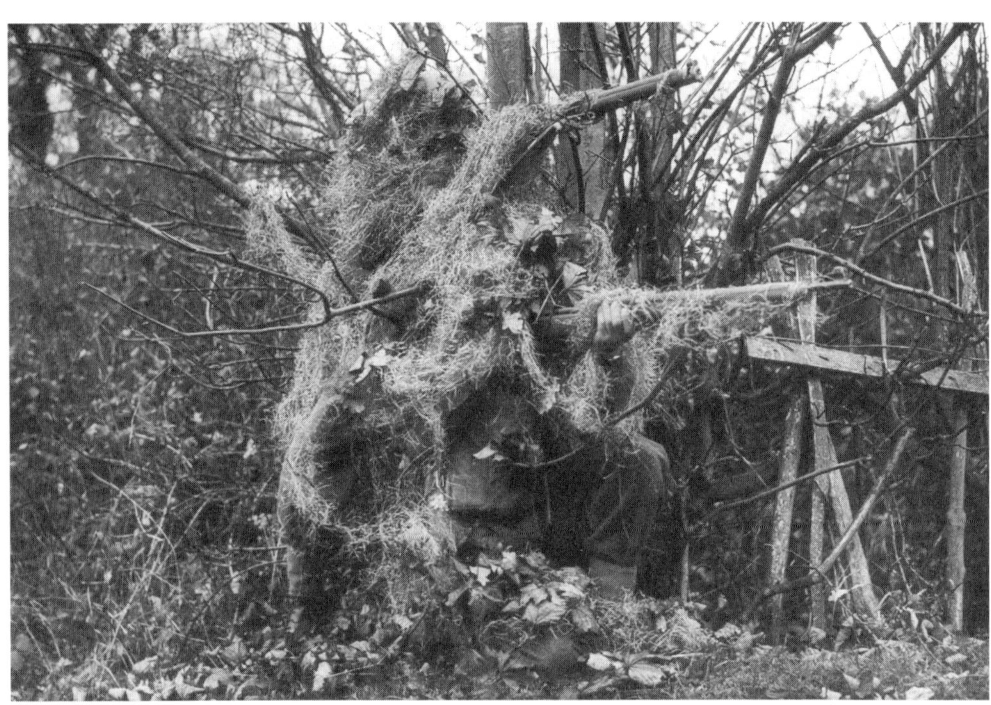

Figure I.4—**Self-concealment**. Two infantrymen use steel wool drapery for camouflage. Official War Office photograph dated November 11, 1940. Reprinted courtesy of the Imperial War Museum [H5464].

model, and reproduce the real-time color change found in certain animals. The ultimate goal becomes a technological approximation of visual evanescence.

The logic of camouflage is the translation of "a seeing into the world," to quote Roger Caillois, into a structured and ordered behavior. This "seeing" is activated precisely by the rendering invisible of the self.[14] Caillois and the psychoanalytic critics he inspired have seen this in terms of a pathological "evacuation" of identity, a self-destructive "psychasthenia."[15] Paul Virilio approaches the concept of camouflage in a similar vein, specifically addressing the relationship between technologies of perception and militarization.[16] Camouflage, in Virilio's account, is emblematic of the dehumanization effected by the encroachment of an all-encompassing military-industrial complex. Virilio sees this as a process that continued over the course the twentieth century, so that by the time of the Vietnam War, "objects and bodies were forgotten as their physiological traces became accessible to a host of new devices."[17] In these accounts, technologies take over. People become increasingly passive as images are accepted as truth, at least in an operational sense.

In *Hide and Seek*, however, live bodies, natural objects, and human actions are exactly what are at stake. They are the locus of active processes of self-fashioning and the substrate of camouflage media practices. The onus is on the human subject to shape the form she takes on, as well as the technologies necessary for that form's production (figure 1.4). Camouflage is thus, in its essence, both creative and productive; it is both a logic and a poetic. What is the configuration of self versus environment that enables one to efface the traces of one's own presence from photographic media of surveillance? It is camouflage consciousness, in which full self-consciousness becomes literal photographic self-analysis.

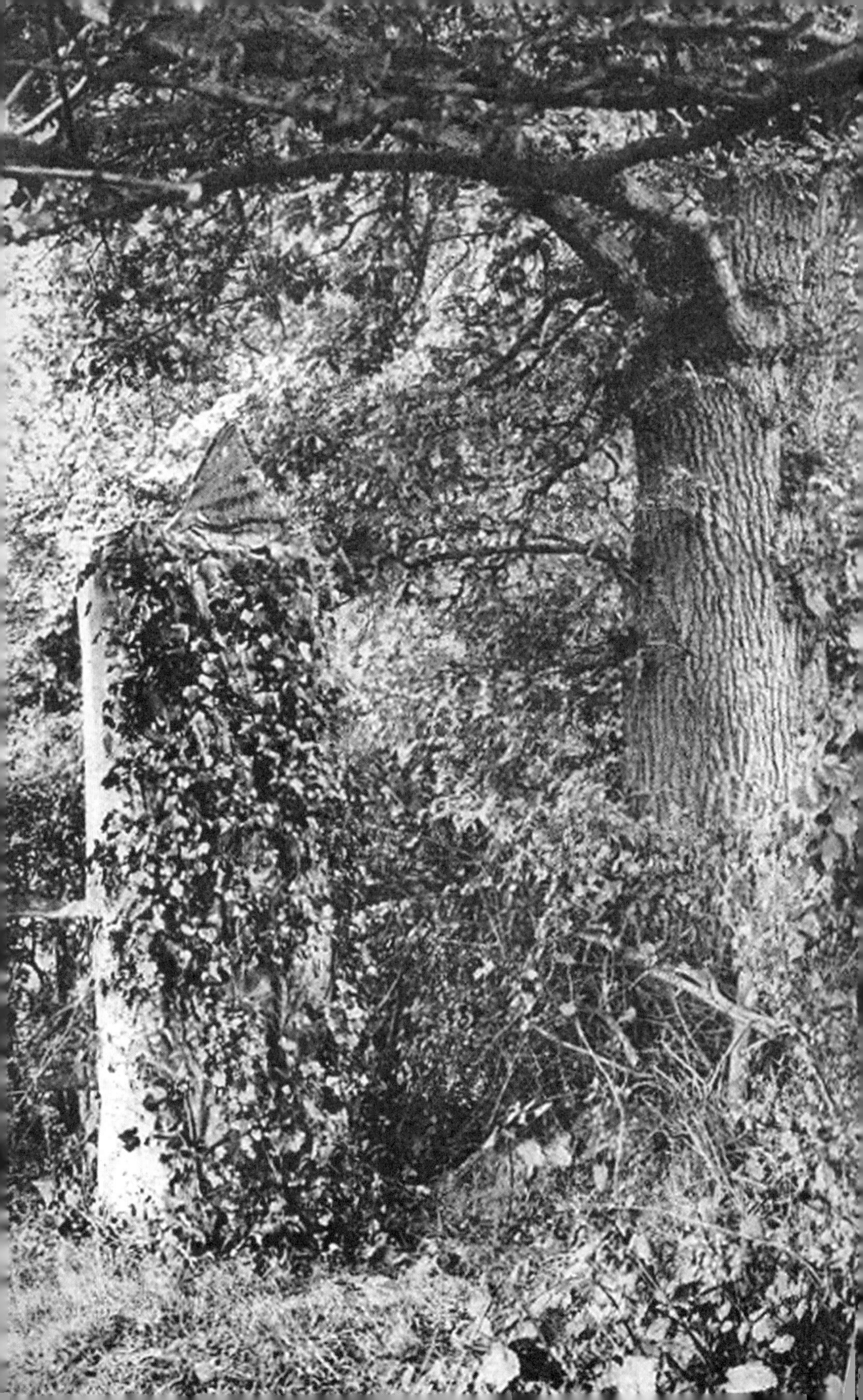

CHAPTER ONE

Productive Mimesis

and the Art of Disappearance

Instantaneous Photography and Abbott Thayer's

Modeling of Invisibility in Nature and Man

On November 11, 1896, an American painter known for demure landscapes and society portraits made an appearance at the annual meeting of the American Ornithologists' Union in Cambridge, Massachusetts. Abbott Thayer arrived at the Harvard Museum of Comparative Zoology carrying a sack of sweet potatoes, oil paints, paintbrushes, and a roll of wire. Thayer's goals for the afternoon were less modest than his materials. He planned to revolutionize evolutionary biology, one demonstration model at a time. He aimed to prove a supposedly universal theory of protective coloration by making things— all kinds of things—disappear.[1]

The morning of November 11 was spent listening to papers on recent bird sightings and conservation efforts in the museum's lecture hall. After lunch, he gave an open-air lecture on the two principles of invisibility in nature that together formed the "law" that he announced in his title, "The Law Which Underlies Protective Coloration."[2] Nature's nonhuman animals are dressed in outfits, he contended, cloaked by skins that have evolved to obliterate visual signs of their own presence. Such obliteration is not permanent, however, he argued; invisibility functions only at "crucial moments" of utmost physical vulnerability.[3]

Call it "snapshot invisibility"—an animal is concealed in space for the duration of the click of a camera shutter. To explain its effectiveness, Thayer identified two distinct visual phenomena: "obliterative countershading" and "disruptive patterning."[4] In the first, the parts of the animal lit brightest by the sun are consistently colored darkest,

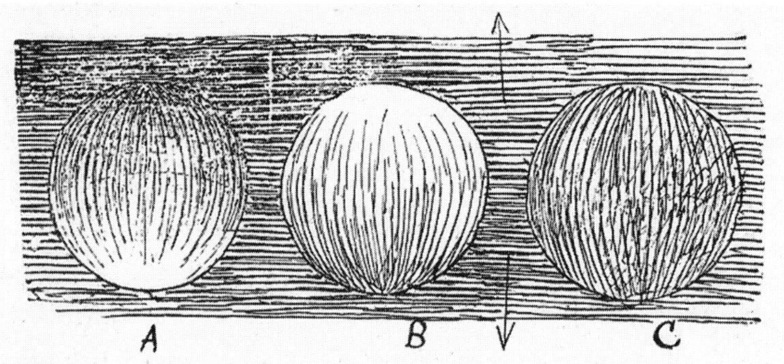

Figure 1.1—Obliterative coloration. According to Thayer's first principle, animals are colored in nature as in A, and lit by the sun as in B; and these effects cancel each other out, as in C. "The result is that their gradation of light and shade, by which opaque solid objects manifest themselves to the eye, is effaced at every point." Reproduced from Abbott Thayer, "The Law Which Underlies Protective Coloration," *The Auk* 13 (April 1896), p. 125.

while those areas generally bathed in shadows are colored lightest.[5] Witness the white bellies of wild rabbits or the silver undersides of sharks; the resulting visual compression of a three-dimensional form produces an illusion of monochrome flatness (figure 1.1). In the second phenomenon, mottled patterns corresponding to the animal's habitat disrupt the contours of its silhouette, resulting in an impression of not being there. One example of such disruptive patterning is the coloration of bullfrogs. Natural selection, continued Thayer, has favored organisms that visually express one or both of these traits; the world brims with momentarily evanescent animal objects (figure 1.2).

Thayer's outdoor lecture was not the culmination of his investigations, however, but merely marked the beginning.[6] In two widely cited publications devoted to animal coloration, "The Law Which Underlies Protective Coloration" (1896) and *Concealing Coloration in the Animal Kingdom: An Exposition of the Laws of Disguise through Color and Pattern* (1909), he proceeded to argue vociferously that all animals are endowed with protective coloration—humans excepted. He included animals typically understood to be highly visible, even showy—creatures such as male peacocks, flamingos, and tropical

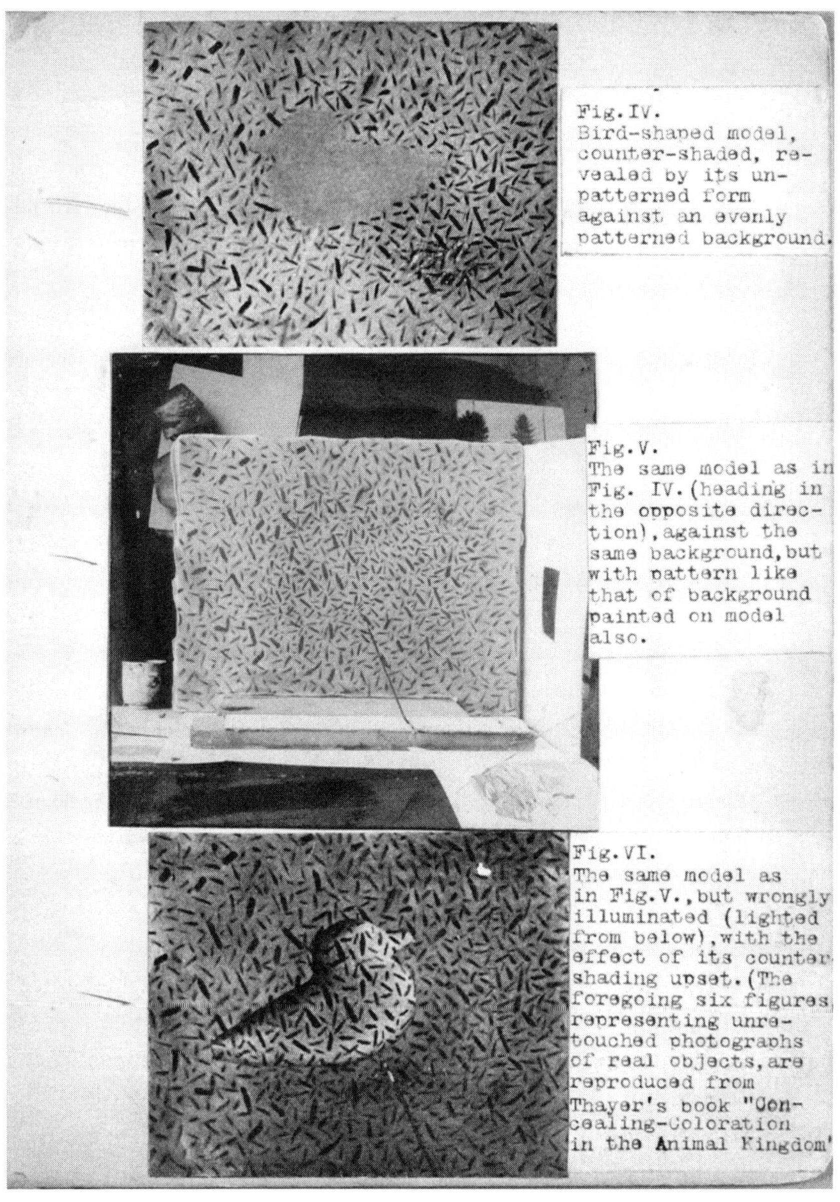

Figure 1.2—Disruptive patterning. Thayer's second law proposed that mottled patterns corresponding to an animal's habitat disrupt the contours of its silhouette, resulting in an impression of absence. Photograph of Thayer's research photographs taken at his Dublin, New Hampshire studio. Reprinted courtesy of the Smithsonian American Art Museum [SAAM 1950.2.35].

fish. That people do not recognize this concealing coloration for what it is could be explained by their failing to see it in terms of a particular moment, against a particular environmental background, and from the vantage point of a particular predator. This was what Thayer referred to as the "crucial moment." At that crucial moment, if you took a photograph, nothing would show up.

For the audience of scientists, bird enthusiasts, and interested passersby assembled at the Museum of Comparative Zoology that November morning, Thayer introduced his law as a greatly important scientific discovery, one uncovered through the workings of an artistic mind. He also performed a highly choreographed game of hide-and-seek. Working with painted and posed sweet potatoes, Thayer made the ones he had painted lighter on the undersides—"countershaded"—disappear from view.[7] Unpainted monochrome sweet potatoes, meanwhile, stood out like sore thumbs against the dirt. "The effect was almost magical," recounted one audience member.[8] A heated discussion about these illusionistic effects, both as witnessed in situ, and as documented in a series of published photographs, lasted well into the evening. A flurry of letters from biologists, bird lovers, and image makers alike soon reached the editor at *The Auk*, the nation's chief ornithology journal.

The excitement derived in part from the context. Thayer's use of models and public performances such as this one, as odd as it might seem from our vantage point, would here and in years to come elicit great interest from celebrated figures toiling in the fields of biology, art, politics, and psychology. The philosopher William James, the biologist and cofounder of evolutionary theory Alfred Russel Wallace, the painter John Singer Sargent, and the English statesman Winston Churchill all hung on Thayer's words at one time or another. Theodore Roosevelt engaged extensively with Thayer's findings shortly after finishing his second term as president. One might assume that Roosevelt would have had better things to do than debate the colors of animals, but the practical, philosophical, and even political implications of Thayer's multidisciplinary speculations were never in doubt, even when contemporaries disputed his conclusions. His ideas, his models, and his theories would in fact have a profound influence on the military practices of the World War I era and beyond.[9] Neither pure science nor aspiring to high art, Thayer's work occupied a vital node in the intersecting genealogies of evolutionary biology, art making, and the techniques of camera surveillance as it steadily encroached upon both military and civilian, public and private life.

Thayer's "law," as illustrated that afternoon, was the strongest refutation to date of what had long been referred to as "mimicry." Mimicry, as famously described by English entomologist Henry Walter Bates and German naturalist Fritz Müller, designates the state of resemblance of the member of one species (the mimic), to that of another (the model). To Thayer, the problem with this concept was one of emphasis; the notion of resemblance between two discrete forms did not capture the actual import of an evolutionary process he understood to be grounded in environmental assimilation: "Mimicry makes an animal appear to be some other thing, whereas this newly discovered law makes him cease to appear to exist at all. The spectator seems to see right through the space really occupied by an opaque animal."[10]

Thayer was soon staging variations of his lecture performance at libraries, town halls, and museums throughout New England and in New York City. He traveled to Europe and presented his potatoes in Florence, Cambridge, Oxford, and London. His collaboration with prominent British biologist Edward Poulton resulted in a publication in *Nature*; this, in turn, coincided with the installation of interactive museum displays at the Oxford Zoology Museum and the London Museum of Natural History.

Interactivity was a key component of Thayer's understanding of invisibility and its manipulation as a force in nature from the beginning. A disappearing exhibit he had installed at the London Museum of Natural History remained there until at least 1925, and it was considered both innovative and exemplary as a piece of museum pedagogy and for being among the first "hands-on" science exhibits to be introduced into a natural history museum context.[11]

The display that Frederic Lucas acquired for the U.S. National Museum in Washington, D.C., provides some clues as to how Thayer's "disappearing exhibits" worked.[12] A countershaded model duck hung suspended in a four-sided glass case. The exhibit was lit from above, as in nature at midday, and the countershaded duck operated on a hand crank. It rotated as the museum visitor turned the handle, so that the transformation from visibility to invisibility occurred at the hands and in the experiential time of the participant-observer.[13] She effected the manifestation of the crucial moment. By 1905, exhibits like the one described by Poulton had been installed at the U.S. National Museum, the Natural History Museum of London, and the Cambridge University Museum of Natural History.[14]

Thayer invited his viewer into a certain kind of relationship with the world: a form of subjectivity in which one simultaneously engaged in perceiving and producing visual evanescence. And in the scientific, aesthetic, technological, and political milieus in which Thayer operated, this hide-and-seek was no parlor-game diversion, but serious business. In articles he would later publish on either side of World War I, Thayer took his theories of animal coloration and attempted to apply them to strategic human ends, as well.[15] Site-specific and self-specific mixed-media practices—Thayer's own and those he proposed to his readers, to political leaders, to everyone who would listen—fashioned a cloak for self-concealment both literal and figurative. Thayer looked both out into nature and inward to the human condition. In so doing, he expressed a conception of the (human) self as an entity to be continually reconditioned in visual relation to its immediate environment.

Skins of Nature and Emulsion

As his work proliferated, Thayer increasingly inserted himself into a long-standing debate over the origins, effectiveness, and pervasiveness of protective concealment in the natural world. After the publication of Charles Darwin's *The Origin of Species* in 1859, animal coloration had become a locus of debate among natural historians, artists, and the lay public. It was not as if no one had noticed that animals can blend in with their backgrounds before. But prior to this period, it was thought that perhaps God had "dropped" them into place just so—"nature by design." By contrast, in the new evolutionary model, such phenomena were explained by a gradual "fitting together" over time (figure 1.3). Evolutionary theories, both Darwin's and those of his colleague Alfred Russel Wallace, presented a range of explanations for animal colors. Darwin emphasized interrelations between the sexes as the cause of the showy coloration found in the males of many species; females choose the more colorful males for mating. Wallace, meanwhile, thought that coloring could be best understood as the result of strictly environmental pressures. He interpreted bright hues and complex patterns as warning signals to potential predators, modes for assimilation in the environment, or mimicry of other, more dangerous species.

Thayer's interest in this debate and in nature's visual illusions had its origins in his hobbies, as well as in his classical training as a painter. He was not only a bird-watcher, but also a hunter, taxidermist, and photographer. He kept a journal of bird sightings in the woods around his New Hampshire home and collected dead specimens for sketching and mounting (figure 1.4). Philosopher-psychologist William James, Thayer's friend, fellow birder—and the father of one of his favorite painting students—discussed the experience of bird-watching in his 1890 *Principles of Psychology*, describing the study of "false perceptions" as critical to understanding sensations related to depth, color, and movement. James also brought to his readers' attention the following anecdote recounted by a colleague: "A sportsman, while shooting woodcock in cover, sees a bird with the size and color of a woodcock...but through the foliage, not having time to see more than that it is a bird of such a size and color, he immediately supplies by inference the other qualities of a woodcock, and is afterwards disgusted to find that he has shot a thrush."[16] James also extended examples drawn from hunting to the world of men at war, with enemies within and without: "as with game, so with enemies, ghosts, and the like."[17]

Traditionally, the locus for identification of a particular bird was

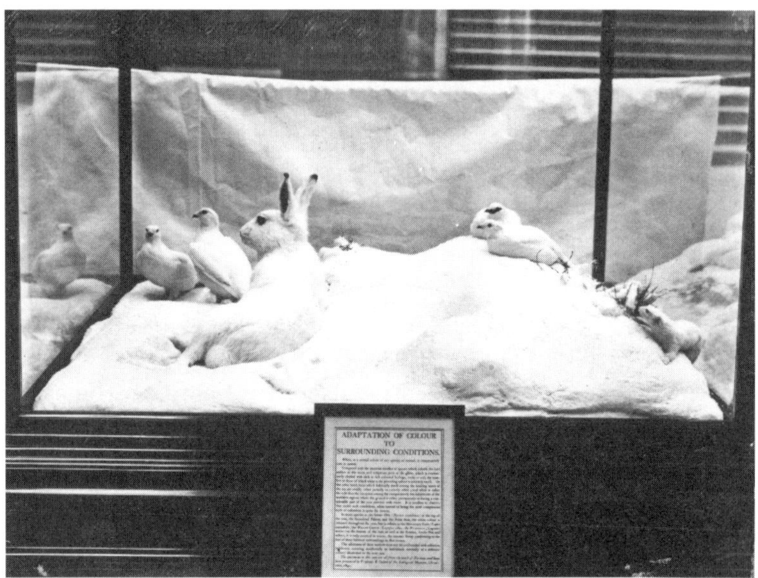

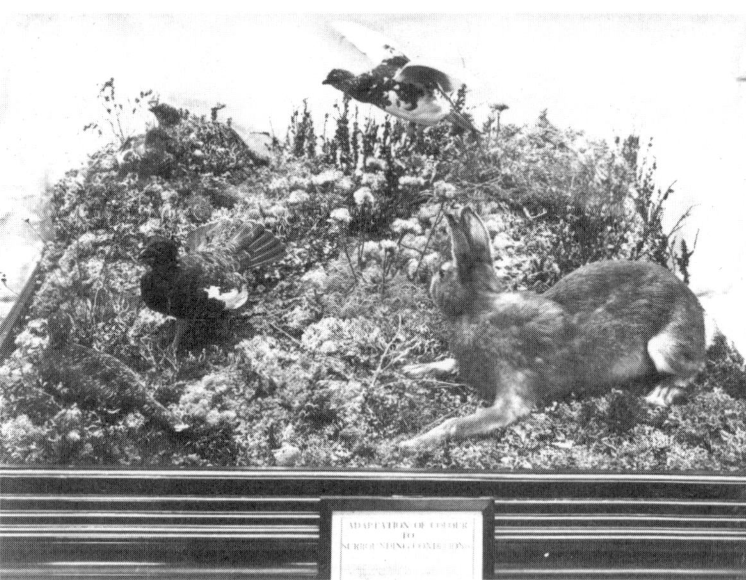

Figure 1.3—Museum groups. Was protective coloration the product of divine design, or the result of an evolutionary response to environmental pressures? Late nineteenth-century habitat groups at the London's Natural History Museum were often organized to demonstrate principles of nature. Top: *Arctic Hares in Natural Environment* from an 1892 exhibition album. Bottom: *Field Animals in Natural Environment*. Photographs by Miss K. Marian Reynolds. Reprinted by permission of the Trustees of the Natural History Museum [NHM PH/173/179–180].

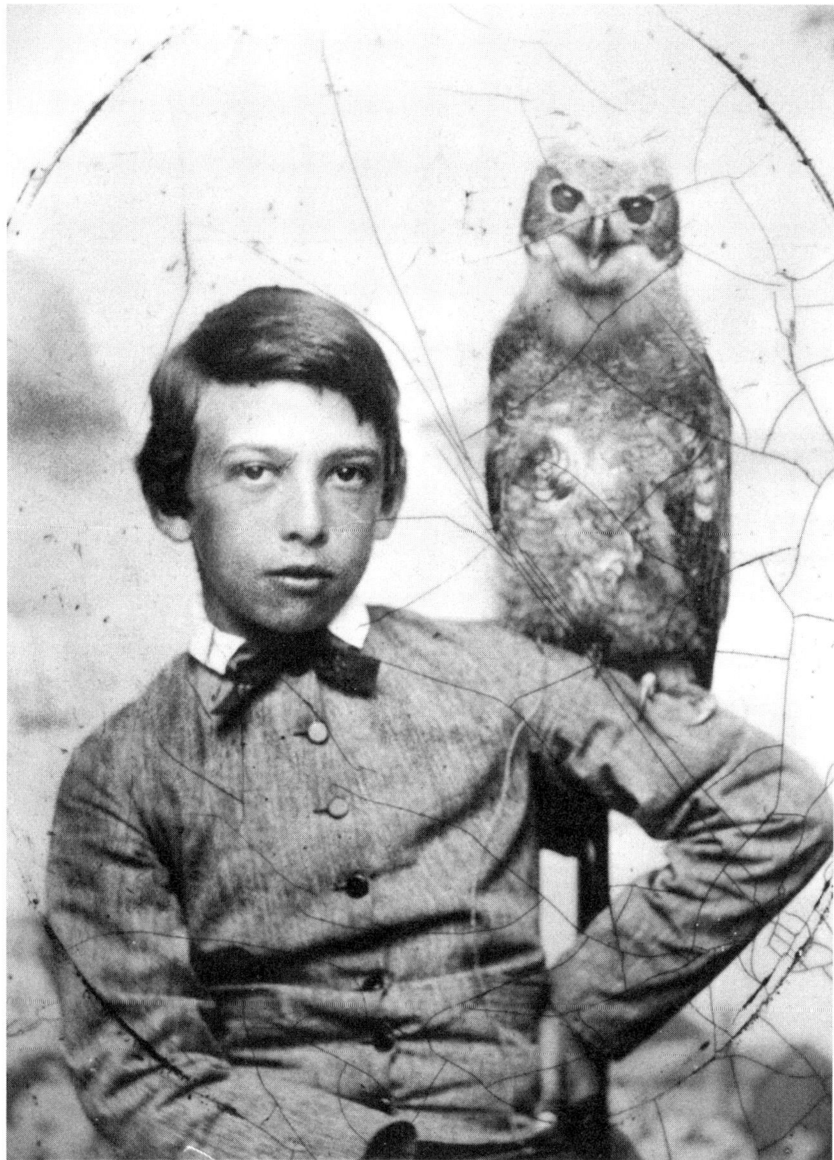

Figure 1.4—Dead or alive? A youthful Abbott Thayer poses with his pet (stuffed) great horned owl, ca. 1861. Photograph reprinted courtesy of the Abbott Handerson Thayer and Thayer Family Papers, Archives of American Art [series 8, box 5, folder 5, image 1].

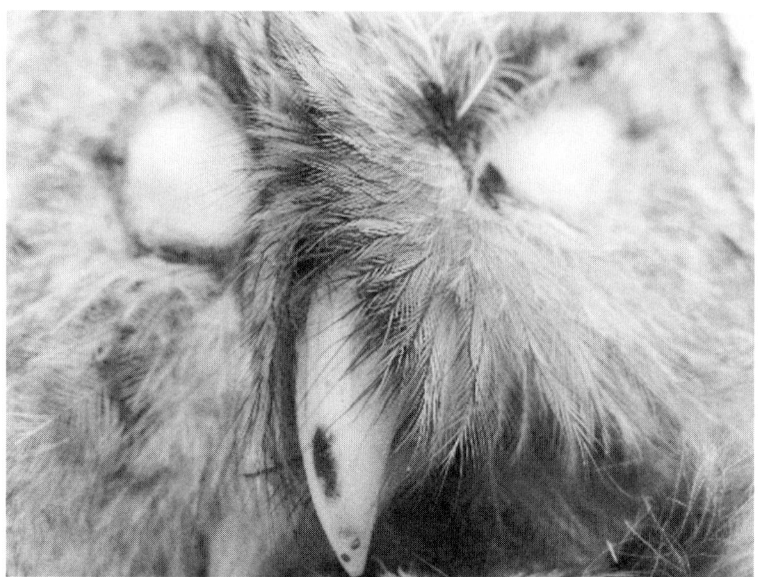

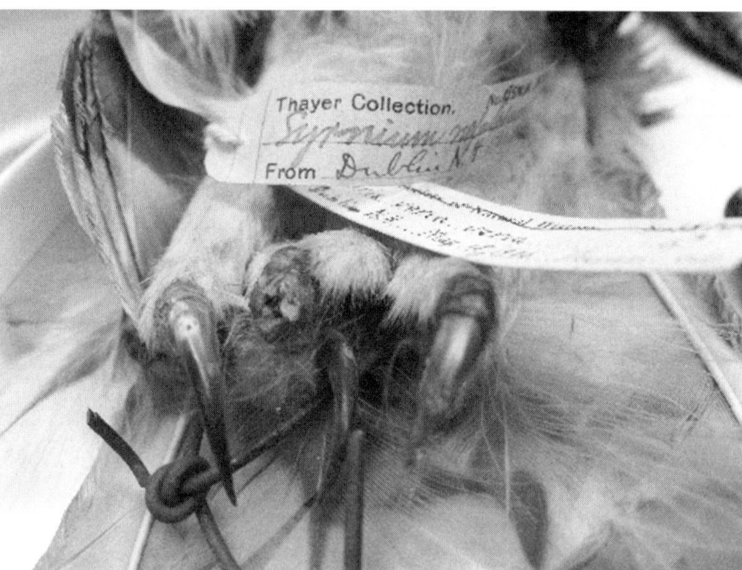

Figure 1.5—Study skins. Taxidermic preservation was a vital step in Thayer's photographic explorations of nature. He hunted, skinned, and preserved animals for future study, including this barred owl (*Strix varia*). He collected this specimen in 1900 and donated it to the Boston Society of Natural History in 1901. The specimen eventually ended up in the ornithology department of Harvard's Museum of Comparative Zoology [MCZ 291744]. Photo by the author, with thanks to the Museum of Comparative Zoology.

its skin. (By "skin," I refer to both the hide itself and the feathers attached.) The skin both marked the animal as a form distinct from its environment and distinguished it as a member of one particular species, rather than another. Hunters relied on color, pattern, and shading cues, as well as on sudden movements, to identify the presence of a bird. Once the bird was dead, closer examination of the same qualities—size, orientation, and the color and patterning of the feathers—aided in confirming species identification (figure 1.5).

Skin specimens also had significant meaning in the context of nineteenth-century nature study. The skin bore an indexical relationship to the animal itself; it was a material trace of a life lived and a clue to the nature of its existence. Amassing bird skins, and animal skins in general, became a way to amass nature, and knowledge thereof, itself. In individual collections, as well as in university and public museums, accumulating skins, classifying and schematizing them, were seen as a necessary part of conquering nature, part of an epistemology, a particular way of knowing based on possession and systemization that dated back to Carl Linnaeus.

Animal skins, collected, identified, and cleaned, also became media technologies. Taxidermy, a form of sculpture in which authentic animal skin is affixed to a molded body shape, became increasingly popular among naturalists in the mid to late nineteenth century.[18] It transformed a flattened two-dimensional skin into a three-dimensional mimetic representation of the living animal. Taxidermied birds, not living ones, became the accepted models for bird art and identification-guide illustrations. James Audubon, in his four-volume *Birds of America* (one of Thayer's favorite books), painted from dead and taxidermied specimens, because they could be examined over a longer period of time—and therefore in much greater detail—than their living counterparts. By the late nineteenth century, the "New School of Taxidermy" endorsed increasingly lifelike mounting techniques, including the use of an excelsior interior "manikin," an artificial bird form made out of a mélange of clay, wire, and twine, around which the original specimen's skin was wrapped.[19]

Three epistemic roles of the skin can be articulated. The first way in which skin transmits knowledge is in its function as the border zone between an organism and its environment, between inside and outside. More plainly put, skin serves as a means of identification, as both boundary and interface. Skins are both how an organism appears as distinct from its milieu and the material by means of which visual illusionism is possible at all. The second epistemic role of skin is as a material entity that serves as a trace or "relic" of the whole living animal: skin as index, as discussed before. This is the sense in which skin is preserved and integrated into collections of bird species housed behind glass in natural history museums. Taken together, the multiple shells or "skins" signify a conquest over nature and a corresponding knowledge claim.

In the third instance, we can consider skin as a medium for technological innovation. Skin may be used to fulfill needs such as clothing or writing parchment. Skins actually produce other things. This is also the sense in which skins are used to create taxidermy sculptures, which exist as simultaneously indexical and iconic in their relation to the once-living animal from which the skins came. Nineteenth-century ornithology practices both drew from and influenced all three of these epistemologies.

This necessarily involved considerable effort and expense. Thayer and other members of the American Ornithologists' Union traveled throughout the country to find their prey. The word "sniper" actually originated as the designation for a hunter who went to means so elaborate and painstaking as to stalk and kill a snipe, a type of bird notoriously hard to track down and even harder to shoot. For those who shied away from the hunt, specimens could be purchased from vendors at open-air bird markets. Thayer traveled all over Europe with his students and son to purchase wild birds. Peregrine falcons and eagles shot in more exotic locales were shipped back home to New England.

Ideally, the taxidermist charged with creating a "life" out of the dead animal "thing" would make the bird look "as good as new" or "as good as living." The taxidermied animal was caught forever in the moment, its skin sliced out of time and posed to appear unchanging for as long as possible. There was a poetry as well as a science behind this art. "The task of the taxidermist, if properly appreciated, is a grave and serious one.... It is to impart to a shapeless skin the exact size, the form, the attitude, the look of life," proclaimed one eminent practitioner. He continued, "It should be an exact copy, as if it were a cast of the animal as fashioned by nature's cunning hand."[20]

But, to the horror of this taxidermist, by the end of the nineteenth century photography began to fulfill many of the same epistemic needs (or Romantic desires) as the killing, skinning, and stuffing of birds did. Of course, photography played a crucial role in transforming nature study more generally. In the 1870s and 1880s, ornithologists quickly perceived that the new portable cameras could stop, so to speak, live animals in their tracks. Increasingly rapid exposures made it easier and easier to capture fleeting moments on film. If bird-watching, and bird hunting, were to some extent games of hide-and-seek, the photographic apparatus became an exciting new player.[21]

Figure 1.6, both parts taken together, expresses the complicated relationship that was developing between people, cameras, guns, and birds in these years. They show the operation of a photographic blind, a structure modified from a traditional hunting blind, built both to conceal a hunter from his game and to provide an effective vantage point for locating his prey. Observers of nature were said to employ a "sharp eye" in the project of imaging invisibility and also when shooting down creatures who tried to effect invisibility, but failed. The metaphor worked with both gun and camera, though only fatally in the former instance. The American physician and inventor Oliver Wendell Holmes (father of the Supreme Court justice) published an essay called "The Stereoscope and the Stereograph" in the *Atlantic Monthly* in 1859 that captured a sense of this vital discursive connection between animal skins and the photographic emulsions that documented them. As Holmes saw it, photography would help humans apprehend nature, and "every conceivable object of Nature and Art will soon scale off its surface for us," he wrote. Looking forward into the later nineteenth century, he continued: "The time will come when a man who wishes to see

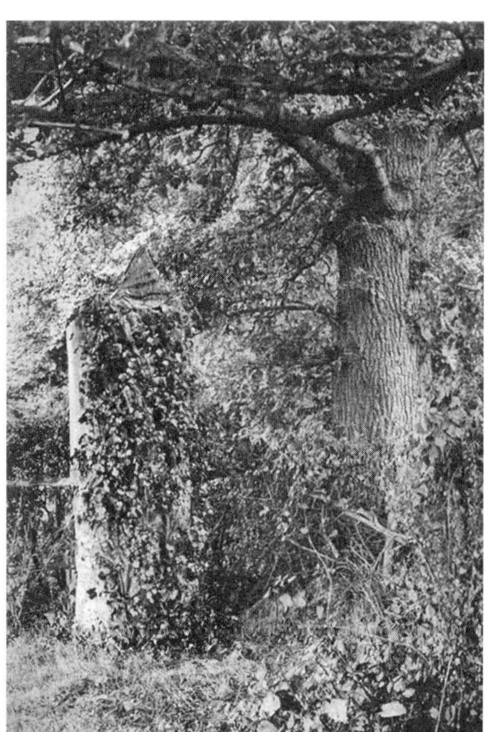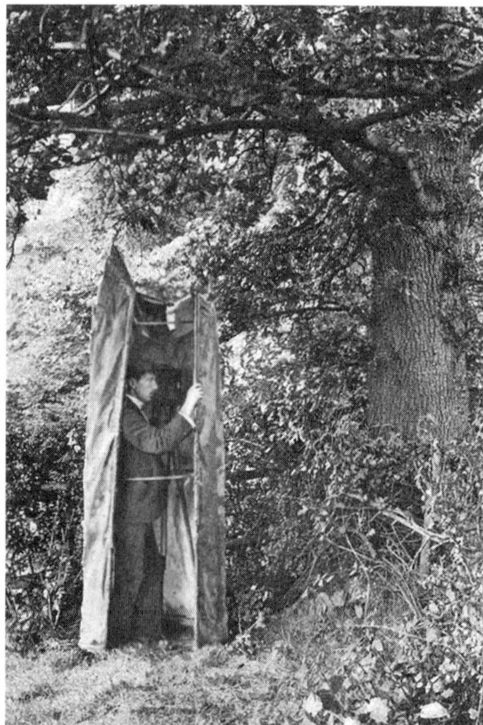

Figure 1.6—Photographic blind. Modified from a traditional hunting blind, this structure is designed to provide bird photographers with an ideal hiding spot from which to see and to shoot. Reproduced from Richard Kearton, *Wildlife at Home: How to Study and Photograph It* (London: Cassell and Company, 1899), pp. 12–13.

any object, natural or artificial, will go to the Imperial, National, or City Stereographic Library and call for its skin or form, as he would for a book at any common library."[22]

Here it is the photographic negative to which Holmes refers as "skin or form." Natural historians, field collectors, and amateur photographers shared in Holmes's vision. Photographs placed in an archive, like pelts in a museum drawer, became documents of the real, accessions to be conserved as knowledge for the future.

By 1894 and 1895, Thayer's taxidermy workshop and backyard had been transformed to accommodate his photographic experiments as well. He had realized that photography, perhaps even more than taxidermy, could be exploited in the interest of discovering, proving, and performing "crucial moments" of protective coloration. Thayer's obsession with invisibility was, like a photograph, specific to a particular instant, snapped out of a continuum of time. Of vulnerable animals he wrote: "When they are on the verge of catching or being caught, sight is the indispensable sense. It is for these moments that their coloration is best adapted, and when looked at from the viewpoint of the enemy or prey as the case may be, proves to be obliterative."[23]

Thayer sought to replicate the experience of looking at this critical moment in the life and death of an animal.[24] From there, he aimed to learn how to *become* an invisible animal, by stuffing animals, painting, posing, and photographing them.

A grouse killed near Thayer's house at the foot of Mount Monadnock helped make the case articulated in "The Law Which Underlies Protective Coloration." After shooting the bird, Thayer brought it inside and attached it to a painter's canvas. Skinning the grouse and removing flesh and bone transformed it from three to two dimensions. As documented in his photograph, this position exposed obliterative countershading face-on. The pattern of feathers was now flattened, disembodied (figure 1.7).

Thayer next mounted these flattened feathers onto the so-called manikin. As his assistant described working with Thayer a few years later:

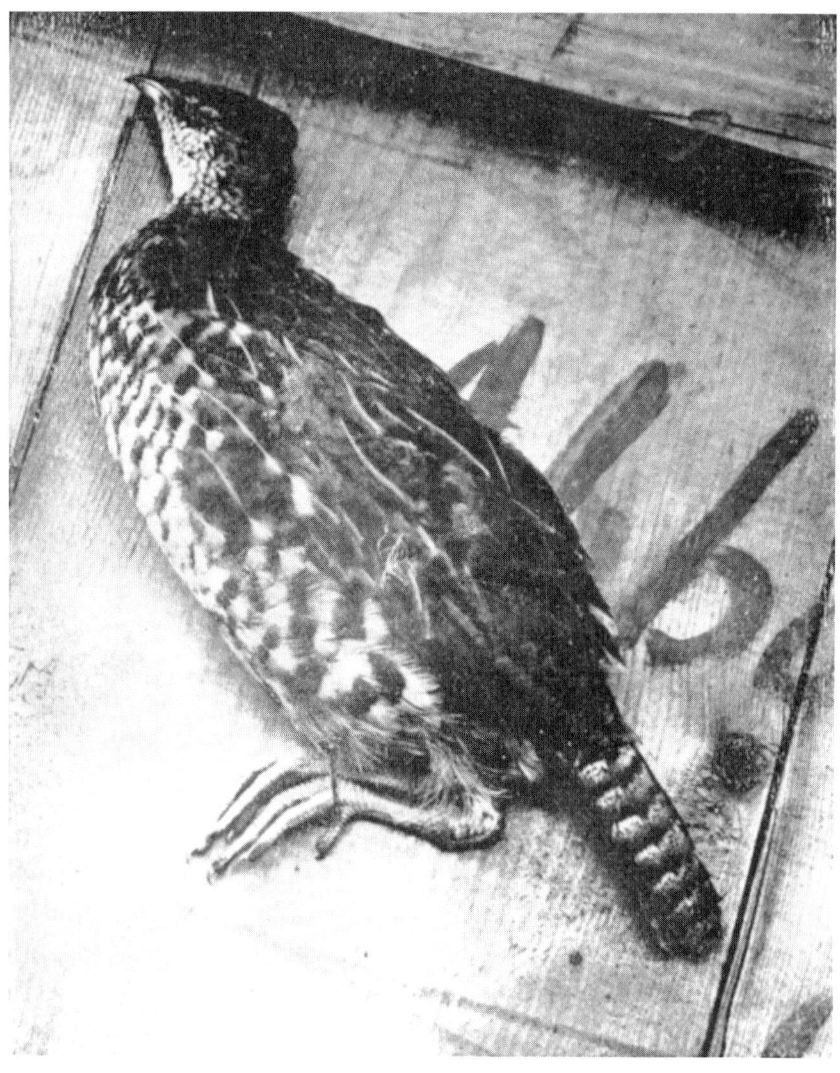

Figure 1.7—Skin as media. This grouse flattened to two dimensions and photographed from the side, served as Thayer's proof of obliterative countershading. The original caption reads: "Side view of a dead grouse to show color gradation." Photograph reproduced from Thayer, "The Law Which Underlies Protective Coloration" (April 1896), p. 126.

> The birds had to be skinned immediately [after collection]…they
> (Thayer and his son) instructed me in the delicate art of enlarging a
> bird's rectum sufficiently to roll its skin tenderly back from its carcass.
> I then cut the carcass away from the skull and wings, put on rubber
> gloves and rubbed the hide with powdered arsenic, rolled the skin back
> in place and stuffed the body with cotton.[25]

The skin was then attached to the wire, bound with hooks, and wrapped around the excelsior form. Each feather needed a cushion as part of the putting "into shape" of the skin.[26]

The laboratory of protective coloration next moved from the studio into the field. Thayer posed the stuffed and mounted grouse in the woods and stepped a few yards straight back; the shutter clicked. Documenting the event in black and white also demonstrated how the grouse, reincarnated as a product of its skin, became all but invisible. Thayer then staged a counterfactual against which to measure his theories and as a means by which to judge the first photograph. For this, he painted the taxidermied bird before placing it against its backdrop. Once posed in the woods, the painted, feathered creature stood out in stark contrast against its background. The camera shutter clicked again. In "The Law Which Underlies Protective Coloration," Thayer arranged these two photographs side by side, both as illustrations and as documents commemorating specific interventions in nature (figure 1.8).[27]

Once in print, the photographs gained the status of surrogate eyes. To use Thayer's words, they were "ocular demonstration" of either static or momentary evanescence—disappearance into the instant of a photograph.[28] For this project, it was necessary that the animals be dead and immobile. Thayer emphasized this last fact when annotating his personal copy of "The Law Which Underlies Protective Coloration." As he wrote in the margins, except for a single photograph of a nesting bird, "all of the others are of course dead and artificially posed."[29] The skin, seen as a piece of the actual animal to which it referred, dead or alive, was an index, a trace that confirmed the relationship of the scene to actual living nature. At the same time, as in the habitat groups that were becoming popular in natural history

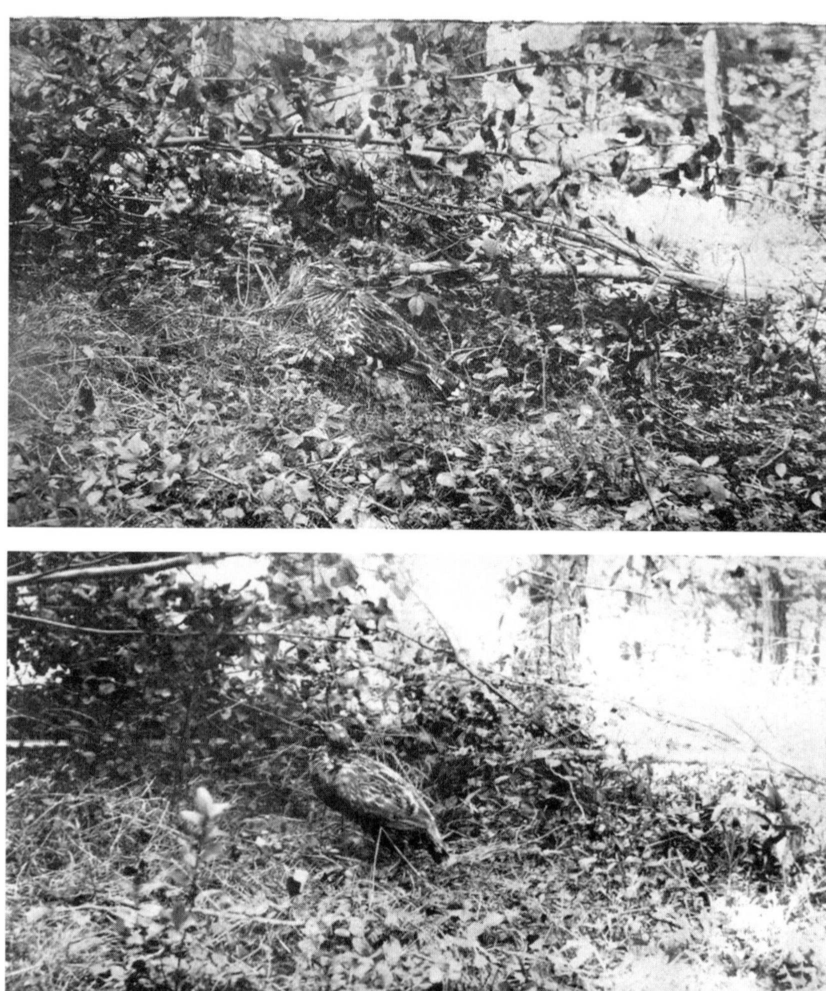

Figure 1.8—Site-specific installations. The taxidermied grouse in top image is posed in Thayer's backyard. The grouse in the bottom image is also posed in Thayer's backyard, though with obliterative countershadowing on its belly painted out. Reproduced from Thayer, "The Law Which Underlies Protective Coloration" (April 1896), p. 128.

museums, taxidermy aspired to render a transitory moment permanent.[30] An instant was idealized through its extraction from any sense of a "before" or an "after."[31] A moment became epidermal, shrouded by its "skin or form," whether taxidermic or photographic in nature.

Thayer also mixed photography and taxidermy in his production of elaborate photomontages—collages composed of photos taken of bird specimens and the habitats in which they'd been killed (figure 1.9). In his workshop, he photographed skinned animals in close-up, cut these images into small pieces, like fabric swatches, and resewed them into a new photographic quilt that reenvisioned the relationship of animal to environment. In a note attached to a photo collage he sent to Alfred Russel Wallace, Thayer described his work as an elucidation of natural process via the production of a photomimetic skin: "Nature has to make, as it were, composite photographs of these animals' backgrounds true only to their average, but in this way they become truth itself—actual art!"[32] The photographic frame—once expanded into the domain of photo collage—became a naturalized staging ground.

To produce this photomontage, Thayer sliced up multiple prints made from a single exposure of the spread wings of a dead great horned owl, incidentally the same species with which he was photographed as a child. Thayer made a separate photograph of the spot at which he had found the owl. He then pasted the wing cutouts onto the photograph of the animal's habitat. In a last step, Thayer photographed his collage. By carving photographs into photographs, he put nature on stage. The artist-naturalist both documented and formulated processes of evanescence in nature, creating a montage of nature in action.

Thayer separately used feathers, themselves, in addition to photographs of feathers, to "paint" the bird's natural habitat. The original bird—a blue jay, say, or a flamingo—became hidden by its own natural clothing. The finished feather painting was a landscape cast in living, albeit deathly, color. Sometimes the feathers were combined with textile pieces, as in one composed of "blue jay feathers against a background of white cotton."[33] In such works, widely exhibited

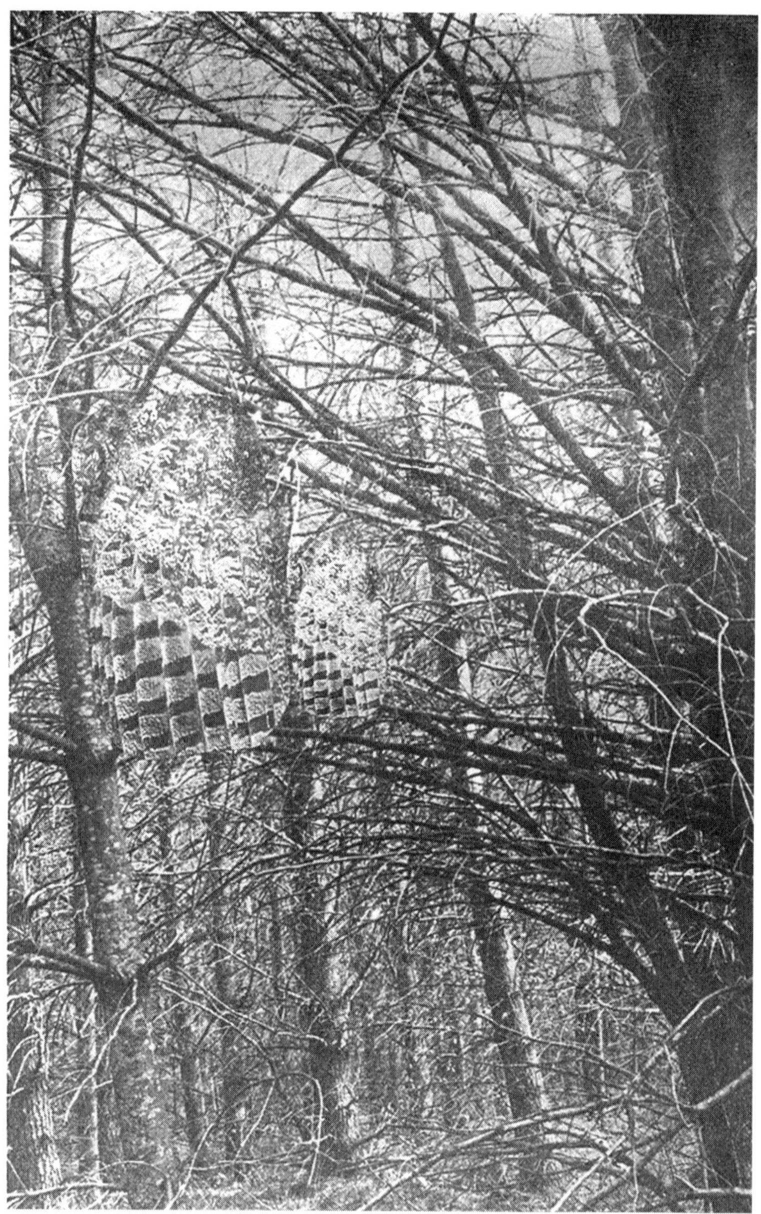

Figure 1.9—Photoquilt. Thayer created this photomontage to show how an owl's patterning approximates that of its typical habitat. Two different sized photographs of the bird's wings are superimposed on a photograph of the location where it had once lived. Reproduced from *Concealing Coloration in the Animal Kingdom* (1909), facing p. 41.

at natural history museums and art galleries and mailed back and forth between biologists in the United States and Britain, Thayer simultaneously produced and documented the processes of a living being's assimilation into its environment. In contrast to the stuffed birds in traditional natural history museum dioramas, these birds were nowhere to be seen. The effect was of a shimmering textile that was also an environmental skin, a skin fabricated and flattened, stretched on a frame and hung on a wall.

Between 1895 and 1905, Thayer sent several of such collages to prominent naturalists in Europe.[34] In 1905, Thayer indicated in a letter that he would enclose a bird of paradise feather painting under separate cover:

> I hope to send you in a few days the bird of paradise sketch made wholly out of this bird's three colors, to show absolutely how he, like probably all conspicuous birds, is an actual picture of his background (in general cases, whatever background they have at the time concealment is worth the most to them), just as my blue jay skin picture shows him to be a picture of the leafless season here.[35]

Thayer then described his "blue jay" feather painting, which Wallace acknowledged in a thank-you letter, as "made wholly of the feathers of a blue jay, or rather of the skin, i.e. patches of it rearranged to show that winter being this bird's exposed time, amidst our deciduous tree-landscape, bare of leaves, he *is* a winter picture of wonderful perfection. In this picture, every detail of the jay's surface (which means, of course, every pattern) appears in full operation."[36] Foreground and background, animal and environment, were mixed. The subject was concealed in its two-dimensional reconfiguration.

Feather paintings of warblers and flamingos followed. They were later exhibited in New York and in London, becoming a permanent fixture of his gallery repertoire.[37] The title "Feathers, No Paint, No Hidden Bird," exhibited in a show at the Pratt Museum in Brooklyn, refers to the fact that the feathers themselves concealed any iconic representation of the bird from which they had derived.[38] The distinct shape of the original animal was not apparent. It was as if, in a materialization of Thayer's theory, the bird became pure background, hidden by the patterns of the traces of its own body.

Stencils and Static Camouflage

Thayer also developed a method of exploring protective coloration—and the necessity for visual skepticism—through the production, installation, and distribution of stencils (figure 1.10). By "visual skepticism," I mean a habit of distrusting the evidence provided by glances and first optical impressions alone. Thayer's stencils were literally cutouts of bird, snake, or human forms. Thayer had begun to create and think through silhouetted animal forms and how they might provide a framework for seeing as early as the early 1890s. By that decade's end, he was proposing their use as teaching tools for visual training. For Thayer, stencils were ways of learning to see, or rather to unsee, into nature.

Is it ever possible to be sure what one is not seeing? Thayer asked this question implicitly with each stencil he produced. The bird artist Louis Fuertes, who had studied with Thayer, recalled his mode of teaching students how to "see" in nature. He recalled: "You have to un-know everything you have always thought and start with a clean slate.... Take a piece of white paper, cut a small round hole in it, and study your color."[39] In essence, Thayer's student had learned from his mentor that to see most clearly what is really *there*, one has to look at it (whatever it may be) through a stencil.

Stencils facilitated interactivity, in this capacity serving as the conceptual heart of his later installation projects. Thayer's interactive tools sometimes took the form of presences and sometimes of absences. In the duck models, a solid form came to seem like an empty space. In the next phase of his work, Thayer tried the opposite tactic. He began with a carefully crafted absence—a stencil—and used it to suggest a presence. A glimpse of the background, be it leaves and branches or grass on which both shadows and sunlight fell, "exposed" the existence of animals, heretofore invisible or presumed invisible, through the stencil's empty space.

Stencils thus became perceptual tools for making real objects both

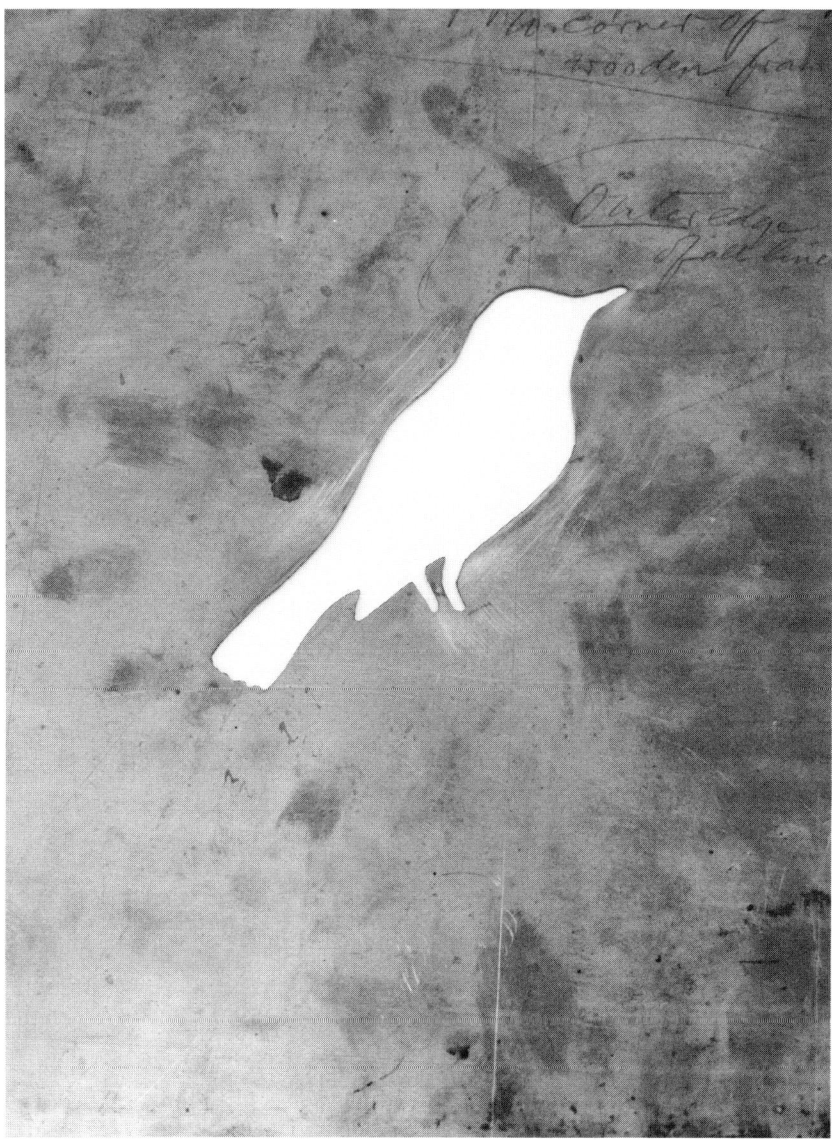

Figure 1.10—Stencil. Literal cutouts of animal silhouettes became a recurring motif of Thayer's investigation of strategic disappearance, as in this stencil of a hooded warbler. Reprinted courtesy of the Smithsonian American Art Museum [SAAM 1950.2.41].

appear and disappear. As part of his study of environmental coloration, Thayer cut the silhouette of a woodland duck out of rigid canvas. Taking wood planks and a tool kit into the field, he then nailed the fabric minus the cut-out area onto the wooden beams attached at crossed angles. He then wedged his construction into the earth at the edge of a streambed. The duck stencil as site-specific installation constructed a generative model of the world as it did or could exist, always with a particular vantage point in mind. Photographs served as proof. The photographs in figure 1.11, side by side, document what an unseen duck would look like (or at least how its skin would appear in a photograph) if it happened to be there and invisible as a result of effective disruptive patterning and obliterative countershading.

Stencil constructions were active models; they modeled a way of seeing the world at the same time that they demanded hands-on user interaction. Thayer advocated the construction and use of homemade stencil sets; they could be cut out of wallpaper or even shoe leather.

He sent kits, including user instructions, to Wallace, to color biologist Edward Poulton, and to artist John Singer Sargent. Construction generally began with Thayer cutting a duck silhouette out of wallpaper. He then pasted the remaining wallpaper to a sheet of watercolor paper, attaching the duck cutout with a dangling string. As he wrote in his instructions to Wallace: "This bit of wallpaper represents any animals' habitual background and explains at a glance how the same habitat can harmonize with very diverse costumes."[40]

Thayer extended the claim in the next letter to Wallace, sent with another wallpaper bird stencil:

All brilliant birds are precisely related to their habitat as the enclosed wallpaper bird is to the wallpaper he fits in showing that there is no such thing, save in a cabinet, as a conspicuous bird. This wallpaper bird, or any so-called conspicuous bird is, even without regard to his background, less conspicuous than if he were monochrome, being cut into separate entities, and when he gets into the place you cut him out of, he is *gone*![41]

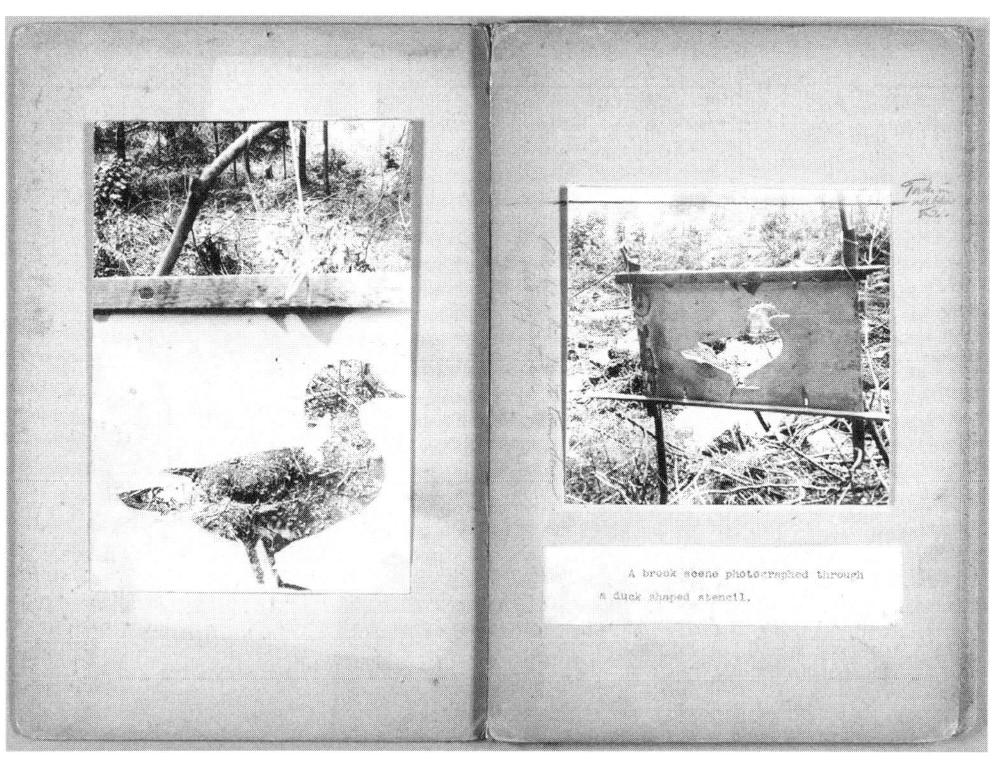

Figure 1.11—Windows on an invisible world. Is this what a duck would look like if it were there but trying to hide? Hypotheticals and woodland duck stencils appear in Thayer's study folder, ca. 1905. Text reads: "A brook scene photographed through a duck shaped stencil." Reprinted courtesy of the Smithsonian American Art Museum [SAAM 1950.2.36].

The duck cutout is a vision and a material instantiation. It is a model of what a duck would look like if it happened to be a perfect wallflower.

Stencils—physical and mental—opened onto other things, but always in relation to a given user at a given point in a field of space and time. The stencil served as visual tool and a field kit in one, instructing in both how to see and how to hide simultaneously. Here, the user was not simply a consumer of illusionistic models, but through the act of consumption produced her own revelation of concealment.

These interactive exhibits and the installations both instantiate what I call "productive mimesis," a process of concrete model building wherein the model is of a natural system that always requires embodied projection into that system. The self enters as creator, viewer, and invisible inhabitant of a remodeled world. Seeing becomes a way of remaking one's own relationship to one's environment.

In the early 1900s, Thayer collaged extant media forms in efforts to prove and better articulate this point. He incorporated stencils, photographs, and cloth into a performative and collaborative form of his original professional métier, oil painting. The canvas became the site for the simultaneous production and representation of the organic world. The stencil kept its central position.

For example, a catalog of Thayer's show at the Pratt Institute Museum in 1924 includes reference to an artwork named *Oil Painting, No Feathers, Hidden Bird Stencil*.[42] Most likely this was an artwork now known as *Hooded Warblers*, currently housed at the Smithsonian American Art Museum. *Hooded Warblers*, produced with the collaboration of his family in about 1908, reveals in material terms the figurative holes, stencils, and dioramic modes of viewing inscribed in the large-scale oil paintings. It is a painting, a field kit, a natural history diorama, and a fine art assemblage in one. *Hooded Warblers* stages a participatory natural history diorama in which (in the absence of the typical standout taxidermy mounts) the viewer is invited inside, simultaneously to hide and seek for her subject. In this capacity, Thayer's viewer, like Thayer himself, is associated primarily with the predator-photographer—with the seeker, more than with the sought.

Hooded Warblers has four separable parts. Short, almost invisible strokes, painted in oils, cover a stretched thirteen-by-fourteen-inch canvas. Taken alone, it would seem to represent the foliage background for a traditional natural history diorama or a sample for a textile pattern or silkscreen print. A seventeen-by-fifteen-inch gold frame encloses the canvas. Inside this frame is a secondary frame, attached to the larger frame with two brass door hinges. The effect is that of an antique museum case or a modern-day medicine cabinet.

The other two parts of *Hooded Warblers* are lodged one atop the other within the cabinet (figure 1.12). Two removable flat panels are made of watercolor paper and copper, respectively. A sheet of ten-by-eight-inch watercolor paper is adorned with representations of two warblers, each life sized; one is painted and the other a silhouette. The painted warbler sits atop a painted tree branch, its style calling to mind the Audubon tradition of bird art with which Thayer was so familiar.[43] Nestled underneath and to the right is the silhouette.

This shadowy doppelganger "cuts a hole" in the white paper, and when placed in the ecology enclosed within the gilded frame, cuts a hole in the foliage as well. The thirteen-by-twelve-inch beaten copper panel lies between the watercolor paper and the canvas. Dead center is the silhouette of a warbler. It is, in effect, a metal version of the canvas "field stencils" that Thayer photographed in his backyard. Transported into the field, the copper panel would reveal both how to hide a bird and where to look for a hooded warbler already hidden. The stacked stencils—paper on metal—reveal a male hooded warbler buried within the foliage background, so that out of a background emerges the very foreground that the background was built to conceal.

Thayer's *Hooded Warblers* functions as natural history model, psychological demonstration, and art piece all in one. Contemporary skeptics, however, were quick to point out that Thayer's model of

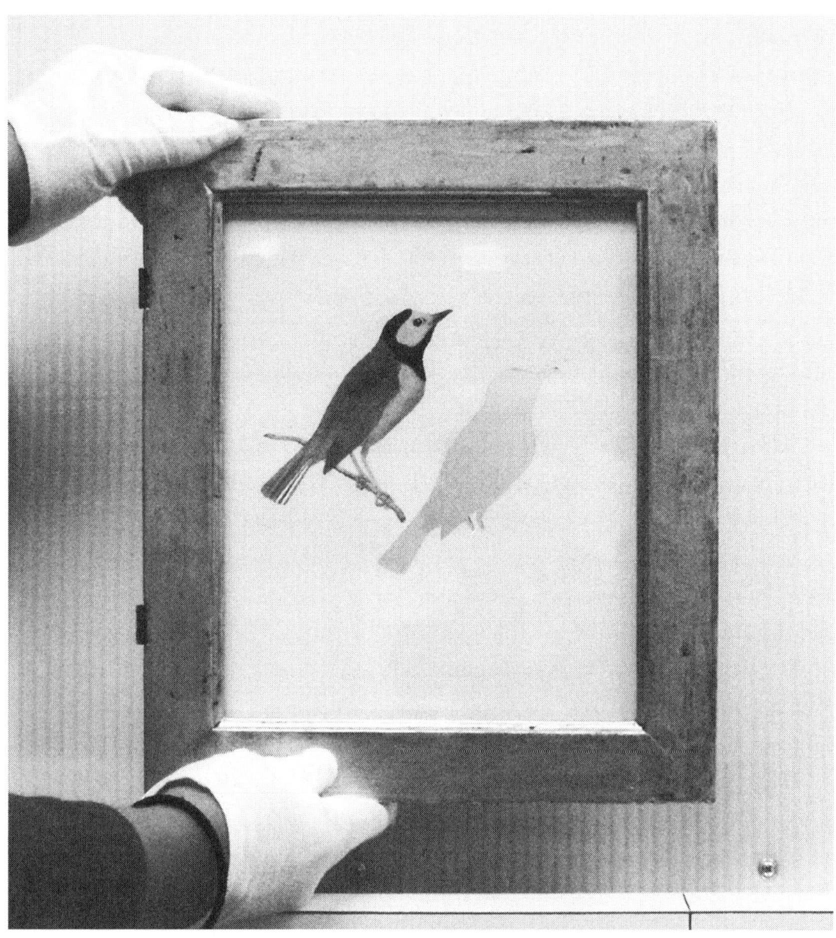

Figure 1.12—*Oil painting, no feathers, hidden bird stencil.* The gold-framed glass "cabinet door" of *Hooded Warblers* is normally attached with hinges to the framed oil-painted "foliage" interior of *Hooded Warblers* (see color plate 16.) When the door is closed, the silhouette cut into the exquisitely rendered natural history painting reveals its doppelganger (see color plate 1).

concealment implied a static environment; organism and habitat alike are of necessity fixed in place. If the warbler shifted positions, or the branches lost their leaves, the illusion of nonpresence would be shattered.

Concealing Coloration in the Animal Kingdom

In *Hooded Warblers*, the stencil is explicitly presented as a guiding motif. In other Thayer paintings, the stencil is implicit in the way in which the painting asks to be viewed. Even if animals didn't always appear to disappear in the real world, they could do so very well in Thayer's assembled universe of paintings, photographs, collages, stencils, and essays.

Between 1901 and 1909, Thayer's family became actively involved in the project. Their New Hampshire home became a year-round laboratory where Thayer also gathered young students, many of whom worked as taxidermy assistants in exchange for painting instruction. Thayer's wife Emma, son Gerald, and daughters Mary and Gladys joined him as fellow investigators, technicians, and artisans. In their backyard lived a rotating cast of animal characters—native species such as ducks, squirrels, and rabbits, as well as exotic imports such as peacocks and copperhead snakes, shipped in from other states.

Each art object produced addressed the enigmas of coloration and invisibility in a different way. In the feather collages, Thayer and Gerald mounted actual bird skins and feathers on panels, calling attention to the animal as pure surface; for others—I call them "photoskins"— they created further photographic collage "quilts" that combined small fragments of larger photos, developing Thayer's idea of nature as a two-dimensional "media environment" where a living body could be made to stand out or disappear as easily as an inanimate pattern. In such photo collages, Thayer isolated the disruptively patterned "holes" on animal skins. Gladys worked on a series of painted assemblages—animals painted in oil and overlaid with metallic stencils.

Thayer and his son collected these various projects in the massive, illustrated 1909 publication, *Concealing Coloration in the Animal Kingdom*. Its frontispiece is *Peacock in the Woods*, the original version of which now hangs in the National Museum of American Art. Thayer and his collaborator, Robert Meryman, based the painting on a taxidermied peacock from the Thayer menagerie. The photographs included in the book were produced as posed *tableaux vivants* of *nature morts*. In these, as in the grouse photos, Thayer created what in the context of a modern-day natural history museum would be called a diorama: a reconstructed moment in a simulated nature. Thayer proudly announced at the outset of the book that any photographs depicting live animals were themselves "gleaned from periodicals or secured by special advertising."[44]

The layout of the book itself, however, expressed a unique approach to visual training; it is organized around large-scale and often interactive figures and inserts. The book's twenty-seven chapters contain 16 full-color plates and 113 black-and-white photographs. Several of the color plates are reproductions of paintings made with the feather collages of blue jays and birds of paradise in mind. Several are painted renditions of the sort of diorama setups whose photographs had appeared in his 1896 publications. The color plate titled *Grouse* is thus described: "The picture of the grouse is a faithful copy of a specimen in a house—lighting artificially arranged to correspond to that which the live bird in the forest would normally have—the background was painted from photos and outdoor color sketches." As in Thayer's earlier reincarnation of a grouse, documented in the photographs reproduced in figure 1.8, both taxidermy and photography were employed to construct a painterly representation of the grouse's visual evanescence in nature (see color plate 3).

The book proceeds from three chapters on the theory of "the universal law" to sections on birds, mammals, fish, reptiles, and insects (including butterflies and moths). Its centerfold is a snake-shaped stencil overlaid on a painting of a copperhead concealed in the foliage (figure 1.13). "This, taken together—paper stencil and painted form—is a *bona fide* study of a Copperhead Snake among dead leaves—its normal situation. So exact, on this snake, is the representation of leaves and the spaces between, that no exaggeration in the painting is possible."[45]

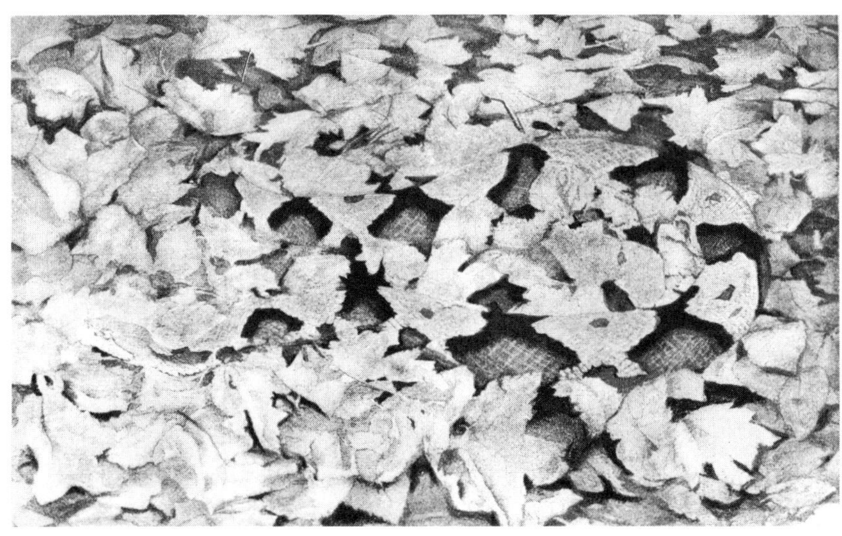

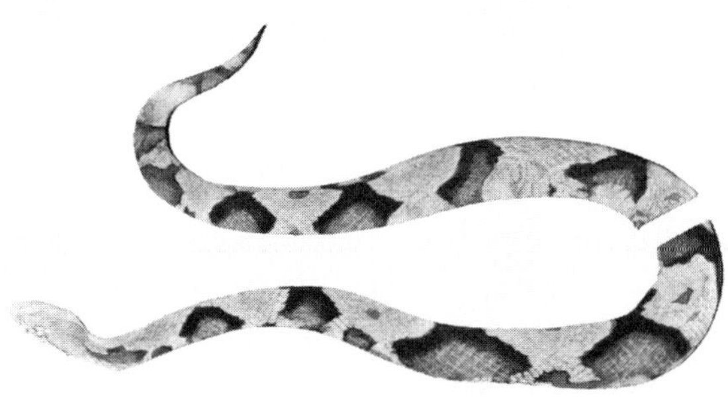

Figure 1.13 — Copperhead centerpiece. Folding a stenciled sheet of paper reveals a copperhead snake (*Agkistrodon contortrix*) smack in the center of Abbott and Gerald Thayer's *Concealing Coloration in the Animal Kingdom* (1909).

Thayer, Gerald, and their friend Rockwell Kent based the painting on extensive studies made of a copperhead on loan to the Thayers' backyard menagerie from the Bronx Zoo. Thayer posed the snake on a perfectly concealing floor of pure New England foliage. In this sense, the snake was not exactly in its "normal situation." But within the habitat enclosed by the bookbinding, there was indeed a whole picture, "taken together," a rendering of the epistemology of static camouflage.

In *Peacock in the Woods* (figure 1.14), reproduced as the frontispiece of *Concealing Coloration*, figure-ground relationships are blurred, and the same colors are used in both the background and in the foreground animal. As Thayer described *Peacock*: "His green-blue head is equipped with a crest which greatly helps it against revealing its contour when it moves. Accompanying its every motion, this crest is, as it were, a bit of background moving with it. The bare white-cheek patch on the other hand 'cuts a hole' like a lighted foliage-vista, in the bird's face."[46] But the painting maintains that the distinction between figure and ground exists at some level; it allows—even encourages—that it be found, not least through the internalized application of a peacock-shaped stencil.

The collaborative *Concealing Coloration* project solidified the relationship between pattern, hole, and silhouette that Thayer had been developing since the 1890s (figure 1.15). Thayer had now articulated in image and words the notion that disruptive patterning is itself a form of stencil making, creating visual "holes" that lead an observer through the body of the animal, as if there were nothing there at all.

It was a technique that garnered Thayer a sizable audience. A broad spectrum of scientific, artistic, and popular journals reviewed *Concealing Coloration*. *The Nation*, the *New York Times*, *Nature*, and *Science* all gave positive notices. Reviewers praised the book for its innovative approach, popular appeal, and more or less scientific basis.[47] British psychologist Edward Titchener, a protégé of Wilhelm Wundt, founder of the first experimental psychology lab in the United States, promoted *Concealing Coloration* as important psychological literature in the *American Journal of Psychology*,[48] and

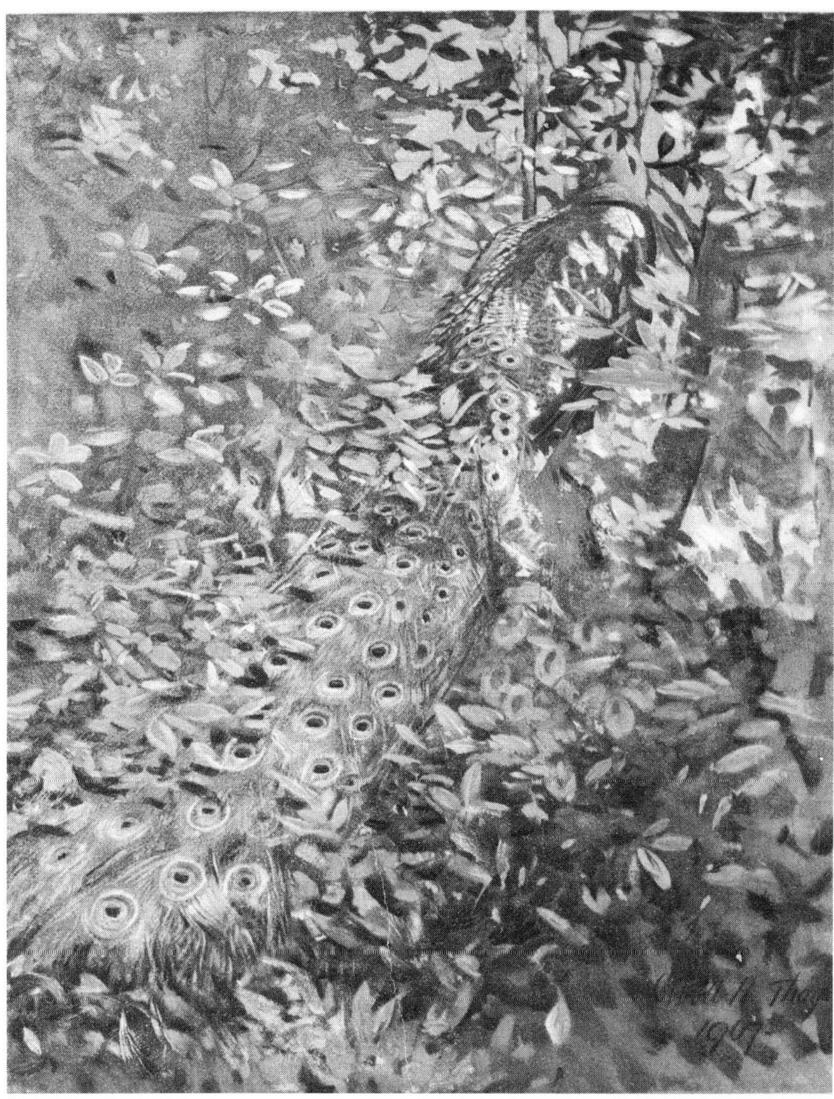

Figure 1.14—Foreground as background. Turning a page of delicately perforated tissue paper reveals the frontispiece to *Concealing Coloration in the Animal Kingdom*, based on *Peacock in the Woods*, an oil painting by Richard Meryman and Abbott Thayer (see color plate 13).

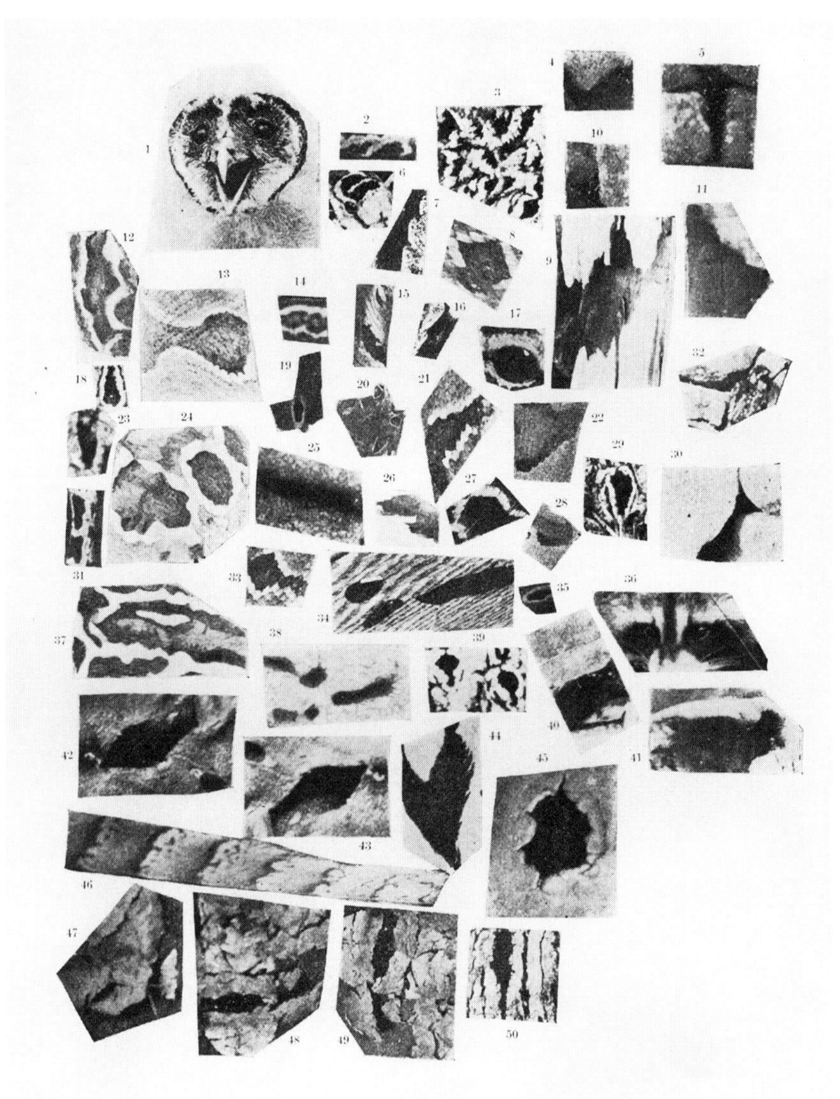

Figure 1.15—Holes in skins. A collage of close-up photographs of the "hole-like" effects. Thayer himself described these photographs as "bits of animals' patterns, all representing holes, i.e. shadowed cavities over which sharply defined edges "relieve" (like wood, leaves, rocks, etc.). Among these are mingled actual holes to show how close is the resemblance." Reprinted from *Concealing Coloration in the Animal Kingdom* (1909), p. 159.

described Thayer's illustrations as valuable optical training devices. For example, "A glance at the plate will train the reader's eye for observation of these facts in the open."[49]

Popular with the public, *Concealing Coloration in the Animal Kingdom* went into a second printing during World War I. But alongside the positive reception and strong sales it enjoyed was considerable argument concerning many of its claims. The reviewer for *The Dial*, a national political and literary journal, criticized Thayer's universalizing approach, though assenting that in this, "its very faults may prove stimulating."[50]

Certain naturalists and politicians, however, took great offense at Thayer's project. The most vociferous criticism came from ex-president and ardent hunter Theodore Roosevelt. In 1910, just over a year after the conclusion of his two-term U.S. presidency, Roosevelt published *African Game Trails*, an extended account of a safari expedition that had begun in 1908.[51] Roosevelt devoted a lengthy appendix to criticism of Thayer's work on animal coloration, lambasting Thayer for extending the principles of concealment in nature to every animal species except his own. Of Thayer's practice, Roosevelt concluded: "such a laboratory is of no more use than a dime museum, and the 'experiments' in it are of very much less use than really good juggling feats."[52]

Though his language may have been extreme, Roosevelt had a valid point. Thayer's conception of the world "out there" as opposed to the world envisioned "in his head" is indeed hard to stomach in its entirety. In Thayer's rendition, protective coloration effects perfect concealment of each individual of an animal species, but this concealment doesn't last all season, or even all day. Rather it pertains only to a single and therefore privileged "crucial moment."[53] The claim was that natural selection favors adaptations that tend to conceal a species against those backgrounds in which it would otherwise be most vulnerable. Evolutionary process, in Thayer's account, promotes perfect—even stencil-like—concealment strategies suited to a given point in space and time. But few animals exist as fixed installations.

For Thayer, each critical moment was a "frame"—here understood in the photographic sense of a frame that inscribes an impression of a given instant of exposure to the reflective conditions of the world, but also in the painterly sense of the wooden structure that supports and encloses canvas construction. Each Thayer painting or construction operated in relation to such an idealized "frame."

According to Thayer, nature's stitches in time are not all equal; some stitches, especially those at which an animal encounters its most vicious predator, matter more than others. Every animal has evolved to disappear in the specific environment and at the specific instant in which it otherwise would have been most vulnerable. The dappled clothing of the peacock, the hot pink of the flamingo, and the yellow splotches of the warbler are each, according to Thayer, associated with a specific instance in time and position in space.

Film theorist Vivian Sobchack reflects that while the stream of time always flows, static photography "freezes and preserves the homogeneous and irreversible *momentum* of this temporal stream into the abstracted, atomized, and secured space of a *moment*."[54] In this abstraction, Thayer's concept of the crucial moment resonates with a deeply teleological bent. The temporal stream in every case is what is extrapolated from the image—it is the set of events both in lived time and evolutionary time. In lived time, it is the getting into position of the animal—what happened before and after the shutter clicks. In evolutionary time, it is understood as a set of bodily appearances (and coloration patterns) led by natural selection in a particular direction. From within this formulation, one can, in fact, frame the nature of the major criticism of Thayer's work.

"What about the moment before and the moment after?"—Roosevelt rightly asked. Altogether, the ex-president devoted 120 pages of *African Game Trails* to refuting Thayer's work line by line, image by image, and object by object. Roosevelt aimed to undermine each of Thayer's oil paintings, stencil assemblages, and countershaded models. According to Roosevelt, Thayer created elaborated versions of nature that he had never seen; for example, he wrote about zebra pelts, and yet had never visited Africa. In an article from the following year published in the *Bulletin of the American Museum of Natural*

History he wrote: "Remember that he has never studied flamingoes in their haunts, he knows nothing personally of their habits or their enemies or their ways of avoiding their enemies."[55] According to the former president, Thayer was a "dangerously extreme member of the protective coloration school."[56]

What made Roosevelt so irate? How did he veer from dismissing Thayer's idiosyncratic installations as mere "juggling feats" to seeing them as the products of a "dangerously extreme" mind?[57] One basis of Roosevelt's distress was his opposition to Darwinian natural selection as the mechanism for evolution; Roosevelt believed instead in something like vague "forces of nature." Another objection was of a more commonsense variety—how, after all, could flamingos be considered protectively colored if you could see them? Roosevelt rejected Thayer's focus on the "crucial moment" as rendered in photographic representation and noted: "the idea set forth in the [zebra] picture is shown to be foolish by a moment's consideration of the fact that neither the oryx nor any other antelope stands motionless at a watering hole. The very fact of coming down to drink implies motion, and motion in such a case instantly takes away all concealing power from any coloration."[58] In Roosevelt's estimation, on the basis of these stylized visual productions, Thayer could avoid thinking about the periods before and after so-called "concealment," on the one hand, and the position of the hunter (in a dynamic relationship with the space around him), on the other.

At issue, too, was fetishization of the photographic instant as a unit of analysis: "As a rule, even the best photograph renders its highest service when treated as material for the best picture," and "we need to consider the relative values of the photograph, the picture and the text."[59] Roosevelt here questions the necessity of the truth value of photographic imagery in an era in which trick photography proliferated. The use of photography in nature study, from Roosevelt's big-game perspective, is at best illustrative and at worst pure theatricality.

Outrage immediately provoked the angst-ridden and self-righteous Thayer to send a series of defensive articles and letters to the *Auk*, which published in the ensuing months responses from ornithologists from all over the country.[60] Then came a heated debate centering on Thayer's use (and perhaps abuse) of photographic evidence.[61] Manipulation was read in multiple ways—as a tool of understanding (as Thayer saw it) and as trickery and deception (as Roosevelt saw it), images "manipulated to fit the author's theories."[62] Whereas it may take only one shot to kill and one shot to make a photograph, the moments that precede and follow the shot cannot be discounted. Roosevelt's argument ultimately won out with the scientific community.

Thayer responded with a series of articles in the *Bulletin of the American Museum of Natural History* and in letters to the editors of the *New York Tribune* and *New York Sun*.[63] A follow-up article in response to Roosevelt's criticism of *Concealing Coloration in the Animal Kingdom* focused on Roosevelt's assertion that forms can't be concealingly colored if you can see them: "No amount of reiterating that you have seen the poacher not poaching or the bank-note counterfeiter not counterfeiting, or this newly discovered animals' costume-scenery-counterfeiter not counterfeiting is any step at all toward finding out whether all three do at certain times perform their tricks."[64] In other words, in Thayer's world, you never really *know* what you may not be seeing. Thayer responded further by producing more models of invisible animals, as well as further essays.

Thayer's critical response also took the form of manipulation of taxidermic, stenciled, and painterly forms. In one photo documentation of a posed "moment," cardboard zebras in striped and monochrome varieties are posed side by side in makeshift rushes. In the photograph, taken—according to Thayer—from the perspective of a lion on the prowl, the striped model is invisible:

> The accompanying photographs are a total answer…a costume such as the zebras…since so perfect a counterfeit of sky and reeds must cause the lion the greatest proportion of failures to notice the zebra when he is still or to keep his outline in sight as he bounds away. To prove that

these sky counterfeits work still better, if possible, in the woods, try your gunny-sack deer and your skunk there, looking at them still from the lower level, as before.[65]

Such mixed-media production and documentation of particular snapshot scenes revealed an epistemology of nature set on disappearance. Thayer finally directly addressed his readers; those experiments deemed "no better than a dime museum" he suggested that everyone construct and install in their own homes:

> Take three pairs of decoys, made of woolen[s], stuffed like a rag-doll and each mounted on a wire pedestal firmly stitched to its back side. Get an artist (or try yourself) to color these as follows. Set one pair of them on very light colored beach sand (or some imitation of such a ground) and color them with pastel all over with the exact tone of this light ground.... By these operations you will find yourself producing delicate sand-colored plovers on the pale sand and on the darker ground...you will evolve a beautiful imitation of some bird like a purple sandpiper.[66]

Remaking nature—in Thayer's language "evolving it"—became an opportunity for mobilizing the perceptual acuity and modeling abilities of the everyday citizen.

Signatures of Invisible Men

Increasing criticism coincided with Thayer's explicit extension of his work from the building of models of nonhuman nature to the realm of articulating plausible human technologies. In cases where he seemed to be reaching beyond the bounds of the conventional understanding of perception in nature, Thayer was laying the groundwork for the application of his ideas to solving engineering problems. In 1912, he drew on his law of protective coloration to explain the invisibility of the iceberg that sank the *Titanic* and killed fifteen hundred people, even suggesting a way to reverse the effects of obliterative countershading and thereby enable the detection of dangerous obstacles at sea.[67] After war broke out in Europe in 1914,

Thayer suggested that disruptive patterns and countershading might be applied to battleships and merchant vessels to great effect.[68]

Thayer soon turned to productive concealment of the human self. He had long been arguing that ultimately, every animal intends to disappear at a crucial moment of vulnerability, and at what moment is a human more vulnerable than during the extreme vulnerability of battle? His investigations into human modes of protective coloration began with armchair anthropologist explorations of tattooing and decoration practices. His "fieldwork" consisted of studying photographs of indigenous people and comparing their appearance to human-shaped stencils cut out of the photographs' backgrounds. Tattooing and decoration practices, Thayer posited, might often be best understood—from an evolutionary point of view—in terms of technologies of protection and invisibility.

Western people, Thayer noted, had tended to avoid such concealing costumes in favor of traditional monochrome dress, as in the case of the gray-clad soldier. This was a mistake, he posited. The twentieth-century soldier, like the primitive warrior, should find a way to "hide in photographs" through an alteration in dress. Thayer pursued these themes in photo collages, in which he dressed people in the photographic skins of the environment within which they are hidden (figure 1.16). His inspiration was "the completeness of an animal's concealing-coloration" which, in his worldview, always "depends upon his wearing samples of all the characteristic details of his background."[69]

Wearing samples of all the characteristic details of one's background is what would soon be called "camouflage," and Thayer made increasingly explicit the relationship between his theorization in art of invisibility as found in nature and his theorization of technologies of invisibility as potentially developed and employed by humans. But even before he made this interest explicit, he had a sense that something more was at stake in his work than either the creation of a mimetic facsimile of nature or the dramatic expression of an argument about evolution. He indicated as much in his note included with the duck stencil kit he sent to John Singer Sargent.

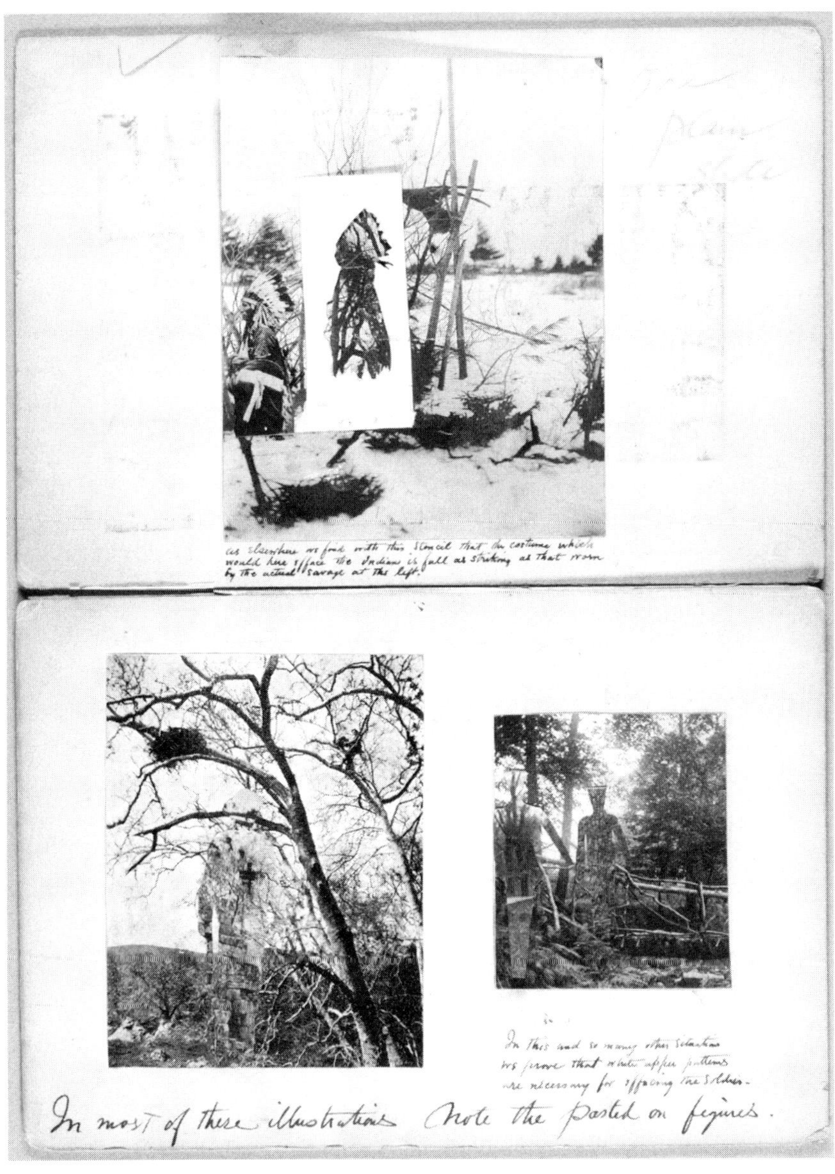

Figure 1.16—Invisible men. Might a human-shaped cutout work? Thayer notes: "As elsewhere we find with the stencil that the costume which would here efface the Indian is full as striking as that worn by the actual savage." In the bottom two images, he presents two "invisible men," both cut out of photographs (one of stone, the other of trees). Reprinted courtesy of the Smithsonian American Art Museum [SAAM 1950.2.69].

Below the duck cutout, Thayer wrote: "to examine the scene in this way through a man shaped hole will tell you what would efface a man from that viewpoint." Before he suggested the literal production of man-shaped stencils (for the purposes of both visual training and brainstorming on military uniform styles), the "man shaped hole" was already on the drawing table, so to speak.

Wallace had highlighted humans' deficiencies in terms of protective coloration in his 1864 essay "On the Origin of Human Races." He remarked first on the adaptation of nonhuman animals toward being "best clothed by nature." In the case of nonhuman animals, Wallace notes: "The animals in any country (those at least which are not dying out) must at each successive period be brought into harmony with the surrounding conditions...keeping exact pace with changes of whatever nature in the surrounding universe."[70] But humans do not "keep pace" as part of this program of natural selection. They resort to technology, instead: "Man under the same circumstances will make himself and build better houses; and the necessity of doing this will react upon his mental and social condition—will advance them while his natural body remains naked as before."[71]

In other words, a human relies on intelligence to create an improved exterior form for herself—in this case, veils of concealment. Man could and can "keep himself in harmony with her [Nature], not by a change in body, but by an advance of mind." From this human need to reshape the visual contours of the body for lack of protective fur or chameleonic scales, then, emerged both the "the capacity of clothing himself, and for making weapons and tools."[72]

Darwin had remarked on the absence of fur as a remarkable and salient feature of humankind; in *The Descent of Man*, he noted that a "most conspicuous difference between man and the lower animals" is "the nakedness of his skin."[73] When the need arose for concealment, whether from cold, from shame, or from visual detection, humans learned to build their own "second skins." The demands on clothing have always been high—armor (protection against shame, enemies, and the elements), aesthetics, comfort, and durability. From their origins in the first days of human culture, textile skins were portable artifacts and temporary prostheses, shaped by the demands of a mobile body and inscribed with markers of that body's history.

Thayer's investigations of tattoos focused on how humans either had or could conceal their natural skins within the framework of static camouflage. In such a framework, bodies were immobile—frozen into the space of a single instantaneous photograph. He determined that tattooing creates effects of "obliterative countershading" and "disruptive patterning" in relation to their wearers' background, and promoted his belief that indigenous tattooing practices were technological adaptations toward strategic self-concealment, despite the cultures' own alternative explanations of tattooing's purpose.

He again adopted stencil making and photo collage as evidentiary tools. Native Americans and indigenous Polynesians became his model subjects. Thayer first created a stencil from the contours of the indigenous figure's silhouette; he then laid the stencil over a photograph of the figure's photographic environment in such a way as to approximate a photographic "mimic" of the photographed human (see color plate 14). The stencil revealed a "cloak," creating a figure where there had been none before; it was the perfect dress for any camouflaged man on the scene. This visual erosion—assimilation of the individual body into the photograph of its habitat—takes place at the site of the skin.

Thayer also assembled photographs of the inhabitants of various environments, none of which he identified. He became an inquirer into these photographic habitats, looking at them through stencils (mental and material), finding locations for the potentially infinite number of concealed human figures therein. Here the photograph provided a more or less accurate mimic of the photographs made of "actual" people. Figure 1.16 (*top*) illustrates his theory about the survival function of tattooing and feather dress among indigenous peoples. A stencil overlay on a photographed environment reveals a figure resembling the photograph made of the alleged warrior.

Humans in a state of nature, then, were also protectively colored, Thayer claimed. But their protective coloration was a result of cultural inheritance and primitive technologies. Individuals framed their aesthetic and practical dress in accordance with an innate conception of how to disappear from view. Strategic concealment, in short, underlay cultural practice.

For Thayer, the world was a media environment, and still photography was the means of determining how to avoid detection. As in the photographic reworking of a Native American in figure 1.17, what is relevant is not who the man is, where he is from, but how he may be understood to exist as a projection of the self—as a figure blending into the given image at hand. Thayer points explicitly to the contrast between indigenous camouflage and the typical dress of the American soldier. Here, the soldier sticks out dramatically.

Thayer began to propose plausible clothing alternatives for the monochrome solider; for the purpose, "the bottom principle is the wearing so characteristic a picture of one's environment" (figure 1.18). Again and again, he asserted that stencils provided the means for locating such a picture. The technique of photo collage underscored the projection of the self into the position of the environment. But exhibited was not merely a visual effect of bodily assimilation. There was also, implicit, a projection of the self, always and already, into an environment that is itself always already photographic. Beyond the level of this projection is the envisioning of how oneself would appear visually within that photograph order. Finally, then, is the necessity of modifying of one's skin to effect photographic assimilation.

In figure 1.16 (*bottom left*), a productive mimetic modeling of a "self" in relation to an environment is revealed. The human subject itself is disguised by an outfit cut from of his own background—stone wall and open air. He takes the form of semitransparent "stone man." Meanwhile the position of the unseeing photograph is also inscribed into the image. A cameraman, diminutive, sits atop the branches in the upper right. He holds a camera apparatus, its lens directed at an enormous, almost pendulous nest. This nest is the apparent target of this photographer's curiosity, or rather its contents are. Perhaps he is awaiting the return of the mother bird. The irony is that—at some level—the depicted photographer is missing the

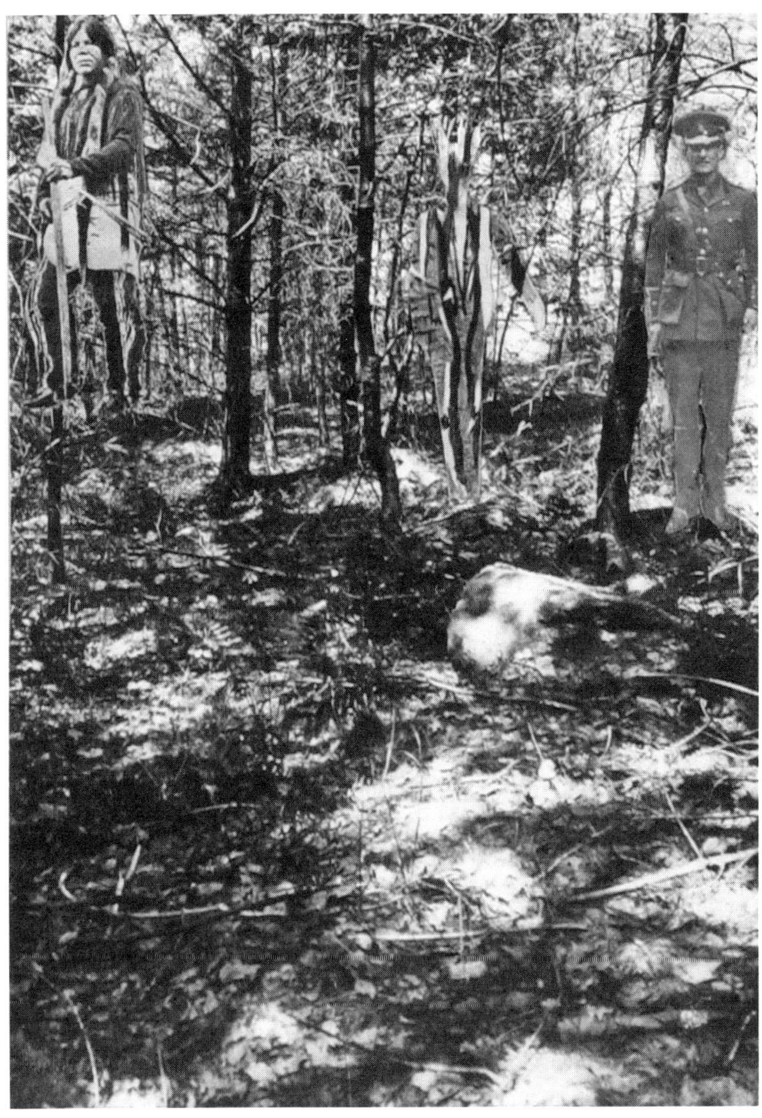

Figure 1.17—Soldier and native. In this photo collage, a uniformed soldier is woefully exposed, while a woman in "native" dress blends into the background. "Note when seen from a few feet away how much more distinguishable the monochrome figure is than the patterned Indian," Thayer added below the collage. Reprinted courtesy of the Smithsonian American Art Museum [SAAM 1950.2.42].

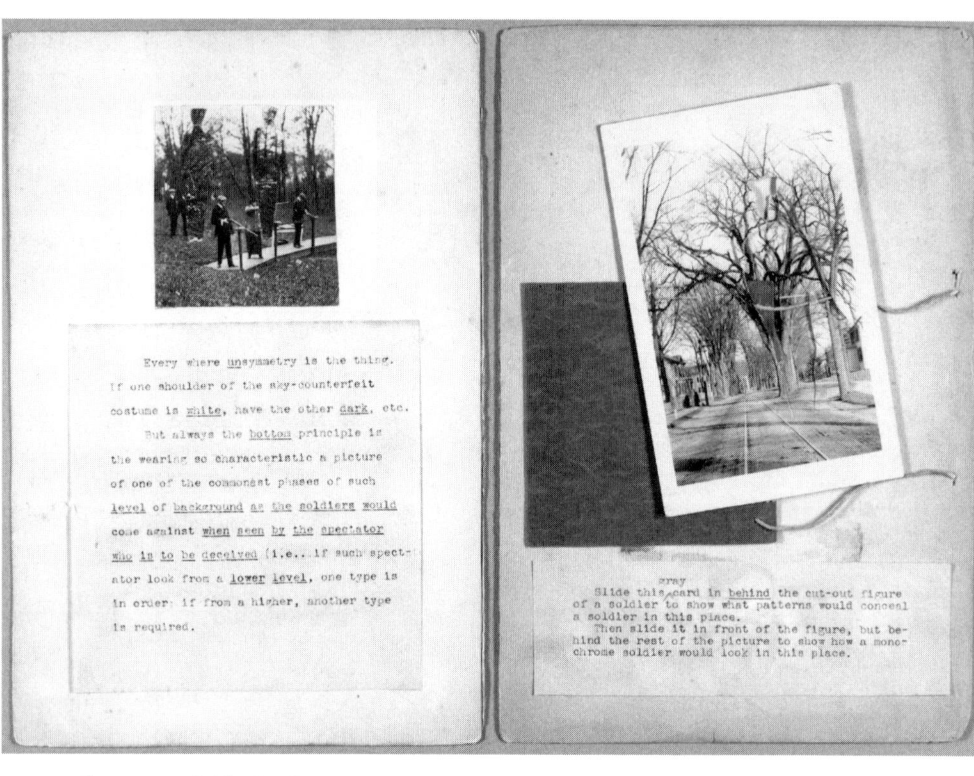

Figure 1.18—Soldiers' uniforms. Thayer's instructions as spelled out in his study folder, ca. 1905: "Slide this gray card in *behind* the cut-out figure of a soldier to show what patterns would conceal a soldier in this place. Then slide it in front of the figure, but behind the rest of the picture to show how a monochrome soldier would look in this place." Reprinted courtesy of the Smithsonian American Art Museum [SAAM 1950.2.63].

point. He looks out at the camera taking the photograph of which he is a part and frames his photograph around the enormous nest and its cozy contents. But he—in contrast here to the viewer—misses the ultimate photographic subject in his midst. If bird-watching is always hide-and-seek, then this photographer is playing the wrong game. The nest, as Thayer has constructed this dioramic environment, is a decoy. It distracts attention from the real concealed subject at hand.

Thayer created the scene behind the scene. He extended its domain from nature to man and back again, using photographic and animal skins as recurrent media. One could say that Thayer's figuration of camouflage in nature inscribed the reconstruction of the human self in relation to nonhuman subjects (and objects) in nature, but also in relation to a regime (imposed environment) of photographic documentation and manipulation. He called into being a version of a world in which forms of self and other and the slippage between the two were made imperceptible. Scissors and glue facilitated the carving of photographs into photographs. There are hidden people and visible people. The image inscribes how to see, how to hide, how to seek, and how not to be seen.

William James's Overcoat

In 1915, Thayer sent versions of these photo collages, along with instructions for their viewing and for the transformation of the views thus witnessed into textile samples, to John Singer Sargent. Sargent had recently been appointed as an official war artist by the British government. Thayer had in fact contacted the British War Office to discuss the mass production of camouflage jackets and pants a year earlier, but his approach had elicited no response. Thayer hoped that Sargent's new position would help him arrange a meeting (figure 1.19).

Thayer's note to Sargent read: "I am enclosing a simple but wonderful device for absolutely ascertaining exactly what any object would have to look like to be wholly invisible against any particular background. You will see at a glance that in this particular picture

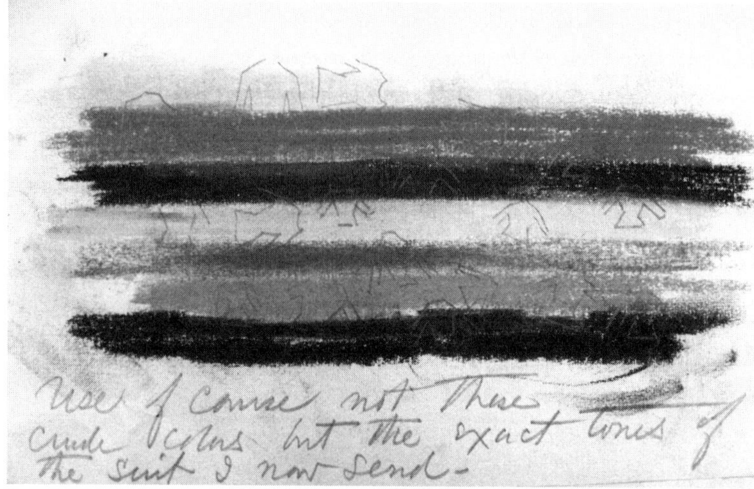

Figure 1.19—Textile pattern. Thayer's sketch for a textile and uniform design, ca. 1915. Oil crayon on paper. Reprinted courtesy of the Abbott Handerson Thayer and Thayer Family Papers, Archives of American Art.

nothing could so entirely efface the soldier I have cut out as for him to wear the pattern which constitutes his portion of this very picture."[74]

Thayer's enclosed sketch for a prototype sniper suit showed a pattern of bright yellow, red, and green splotches onto which he had traced silhouettes of trousers and overcoats. His model for such a jacket, which he went on to propose as a new uniform for the British infantry, was a secondhand hunting coat gifted posthumously by William James in 1910. Thayer was known among acquaintances for the odd assortment of secondhand jackets and woolens that he had acquired and modified over the years, and he wore James's hand-me-down constantly; it became his sartorial second skin, practical for painting in the studio and trekking through the outdoors alike. The coat quickly grew threadbare and acquired a patina, and James's son later recalled with pleasure witnessing "this familiar garment on the back of my dear old 'Uncle' Abbott" and watching it "grow shabbier and more and more covered with paint as time went on."

Thayer attached pieces of dyed and painted rags and fabric swatches to the old overcoat, breaking up the outline of his own silhouette and making him an undetectable presence in the bushes behind his own New Hampshire home. He wore it in his workshop as he composed responses to the criticisms he received for *Concealing Coloration* and began a letter-writing campaign to members of the American government and British Army. At last, with Sargent's help, he arranged a trip to London to meet with Winston Churchill, then the British secretary of war, to show the politician his work.

But once in London, Thayer grew increasingly anxious about the upcoming meeting and paced nervously about the city. His nerves began to get the best of him; it was as if he were hoping to disappear inside something like one of his own camouflage apparatuses. Roosevelt would have thought so, at some level sensing that any man obsessed with asserting the omnipresence of invisible animals must in fact be afraid of facing himself.[75] Thayer, in a state of deeper and deeper agitation, contacted Sargent and told him that he could not go through with the meeting. Recalled James's son, "he [Thayer] sailed for New York leaving a note for Sargent and a suitcase full of demonstration material."[76]

Sargent agreed to deliver the suitcase that Thayer had brought with him for the presentation, which, in addition to examples of painted fabric, collages, wallpaper swatches, and stencil sets, also contained several letters with suggestions on camouflage design. Thayer included a letter commending the British for using khaki suits, but suggested further modification was necessary. Disruptive patterning would be helpful, at the very least, so as to break up the outline of individual human forms.

The reaction was not what Thayer had anticipated. From his perspective, it was a Great Work; from the War Office's perspective, it was a pile of rags. Sargent recalled that "it contained some drawings and an old spotted brown jacket with rags pinned to it"—James's overcoat as abused by Thayer. Years later, the philosopher's son reflected to Thayer's biographer that "this was evidently the last appearance of my father's brown coat."[77]

Like the coat, many of Thayer's experimental assemblages and artworks of effacement have vanished from the material record, leaving behind only textual traces. The duck models, feather paintings, and rag-doll decoys were scattered far and wide and eventually lost. But Thayer's premise of universal protective concealment in nature and its curiously overinductive method eventually did become a pedagogical method for camouflage training and implementation during World War I and World War II.[78] But official affirmation took some time.

The painter Barry Faulkner, former taxidermy assistant, student, and old family friend of Thayer, wrote from New York in 1917: "Uncle Abbott's theories have been used to an amazing extent by the English and French in concealing guns, buildings and every conceivable thing. I've got hold of some of the material and sent it to high personages in Washington.... In case of war this concealing work has real importance and artists are the best people to do it."[79] Alas, American, British, and French military officials did not agree, favoring the schemes proposed by those from an engineering background. Nonetheless, Faulkner and Homer Saint-Gaudens, who also had studied with Thayer, both ended up fighting with the American Expeditionary Force, inspired by some of Thayer's principles for concealment of trenches, supply depots, and infantry at rest.[80]

In his waning years, Thayer submitted several other unsuccessful proposals to the British and American War Offices while continuing to write and paint between increasingly frequent spells of nervous exhaustion. He also contacted Franklin Delano Roosevelt and other members of the United States Naval Board directly.[81] Trials of his sniper suit were allegedly made just as other artists' peculiar prototypes also began rolling in to the British War Office. Thayer died on May 29, 1921, the day after he delivered a lecture on camouflage dioramas he had designed with his son—his last noted disappearance.

Although his proposals failed, his teachings did not. Thayer's students, including Homer Saint-Gaudens and Barry Faulkner, were among the founders of the American Camouflage Society in 1916.[82] Organized in Greenwich Village, it was soon affiliated with the U.S. Army and attached to Columbia Camp and Plattsburg, both camouflage training facilities in full swing by 1917. These organizations' publications and memoranda repeatedly refer to Thayer.[83]

Thayer's works were not clearly science, not clearly art, not clearly models for the manipulation of human environments. In his multimedia constructions of pigment, canvas, metal, feather, wood, and photography, Thayer manipulated tools, media, and disciplinary frameworks. Thayer's work moved between the art studio, the demonstration hall, the field of nature, and the science museum, though never sitting still perfectly well or for long. This precluded the scientific validation he always craved. Despite their encouragement of his ventures, Poulton, Wallace, and the great color biologists of the twentieth century could not abide by Thayer's premise of universal protective concealment in nature or his disregard of sexual selection and his curious, overly inductive method.[84]

But Thayer engendered a new technological form in the process of his overextension. For him, the space of the canvas broadly construed became a laboratory for the generation and testing of new technologies of concealment. As multimedia constructions of skin and pigment, cloth and canvas, fur and feather, his products drew on the simultaneity of presence and absence in the material world. Within this larger political and cultural frame, Thayer and his multimedia studies of protective coloration together occupied a vital node in the intersecting genealogies of biology, painting, and the accelerating techniques of photographic surveillance as they invaded various walks of civilian and military life.

As the twentieth century began, new modes of interaction with natural and man-made objects were springing into being. Seeing and not being seen were becoming coconstitutive elements of a selfhood based on an embodied projection into one's particular filmic and natural environments. In anticipation, Thayer had spun a vital form of seeing and knowing through making—a nascent art, science, and craft founded on the performance of self-effacement.

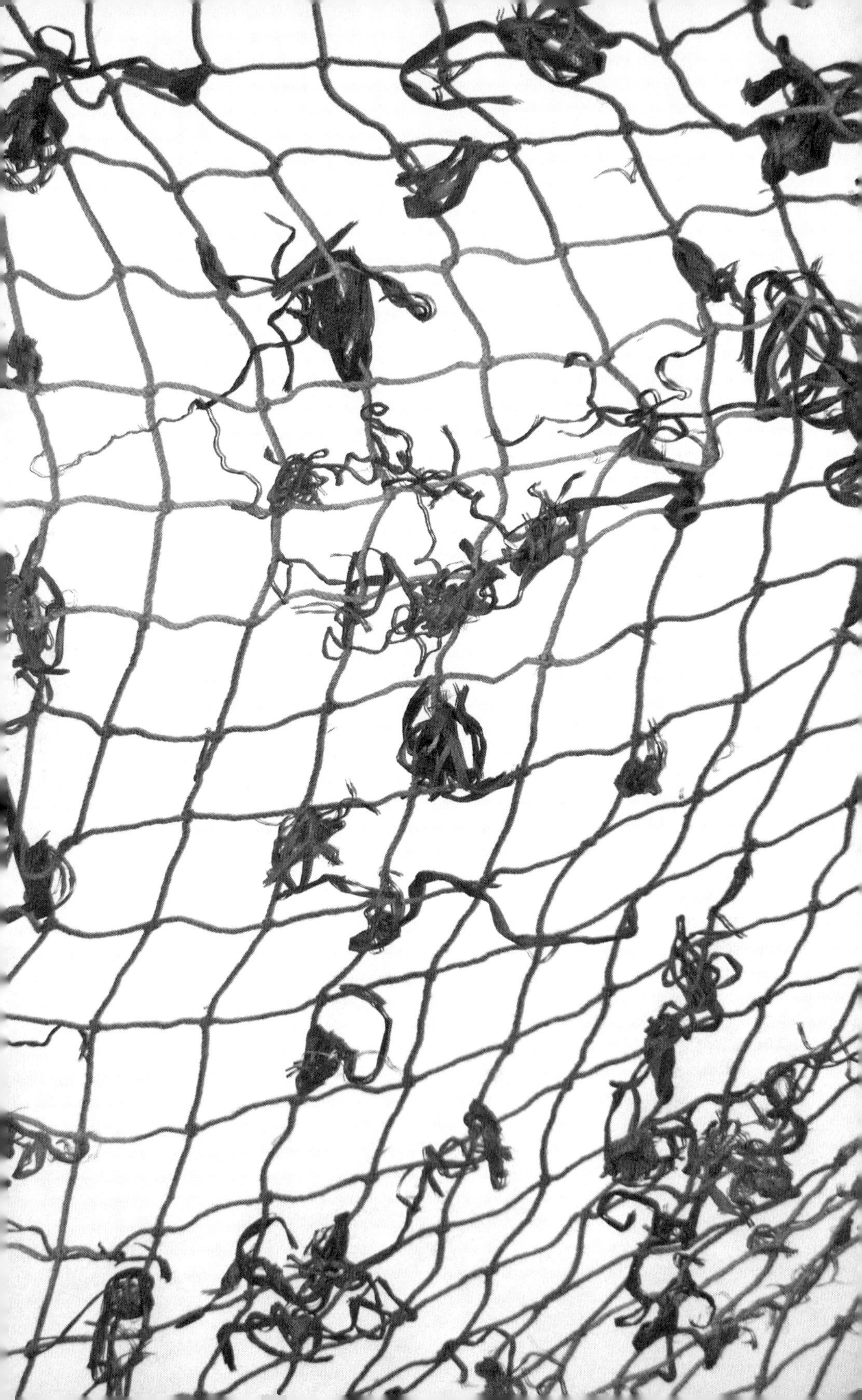

CHAPTER TWO

Mending the Net

Camouflage, Serial Photography, and the Suture of Self-Effacement and Reconnaissance

A photograph album at the Imperial War Museum in London commemorates two years spent at an organization founded to advance the development of "special works," the code name the British War Office had given to camouflage. Winifred Mottershead was the secretary at the Special Works School, and to the front of her commemorative album she affixed the following inscription in precise calligraphy: "The following photographs may help to convey some idea of the development of aeroplane observation and photography during the late War, which created the need for a new science—that of concealment from the air."[1]

Between 1917 and 1919, a motley band of photographers, painters, sculptors, set builders, seamstresses, and engineers trudged to the school, located in a corner of Kensington Gardens, to aid this effort. One great success of their and others' experimentation was netting—the same material heretofore used mainly to catch fish. The Special Works School was both a secret and a showpiece, a site of revolutionary practice and intense speculation (figure 2.1).

The first photographs in the Mottershead album sketch its institutional and geographic topography. Group portraits depict Ms. Mottershead, staff officers, Artists' Rifles' members, and unnamed seamstresses and set builders. These are the *camoufleurs*, men and women drawn from various walks of life and lined up in rows. The photograph captioned "General View of the Camouflage School Grounds" depicts what appears to be a normal park. "Billets at the Camouflage School" shows model military barracks painted to look

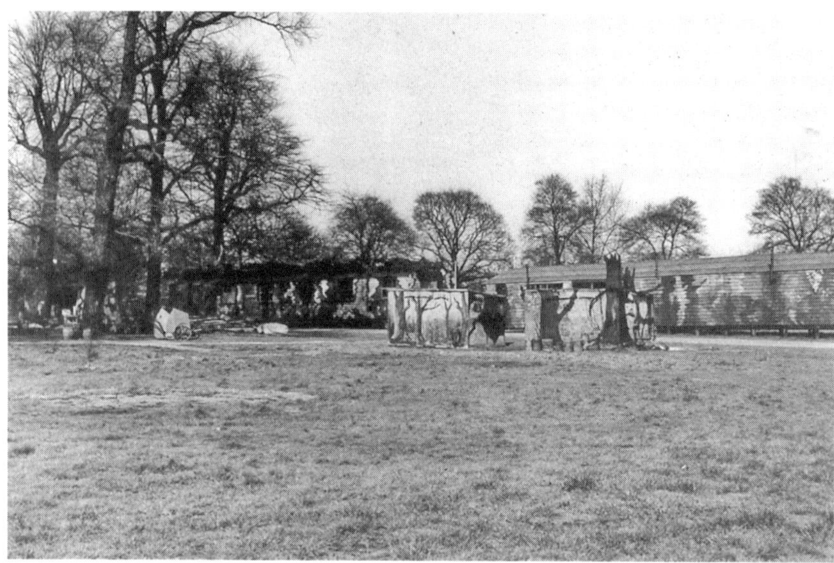

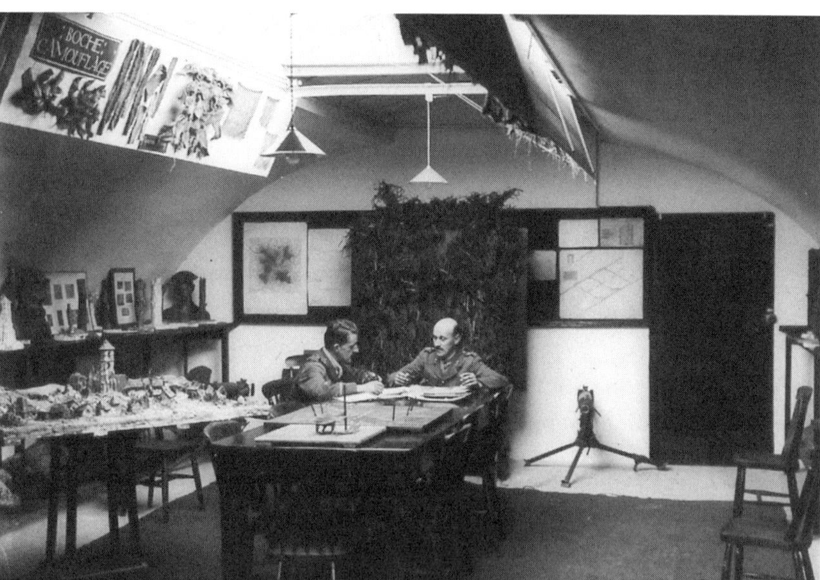

Figure 2.1—Camouflage school. At the Special Works School in Kensington Gardens, London, members of the Artists' Rifles group and other *camoufleurs* attempted to perfect a "new science." Reprinted courtesy of the Imperial War Museum [Q95933 (top) and Q17765 (bottom)].

like local flora—a bas relief of maples, oaks, and birches made of wire, netting, and painted canvas.[2] "Camouflage School from the Air: July 1918" is a shot taken to test the effectiveness of their latest attempts at aerial concealment. The question implied by the photograph is implicit, but unmistakable: Had they met their objective? Had they achieved an adequate measure of invisibility?

These photographs have a gravitas that would seem to befit a military laboratory devoted to a "new science," to use Mottershead's term. But what is one to make of the image, farther on, of a cardboard howitzer partially dressed in fishnet stockings?[3] The photograph seems more like a whimsical production still from Jean Renoir's war film *La grande illusion* than an official wartime document.[4] Photographs further into the hefty Mottershead album proceed to present an increasingly elusive story. To a modern viewer accustomed to thinking of camouflage in terms of two-dimensional splotches of dingy, mass-produced bcige and olive drab, the activities of the Kensington *camoufleurs* seem strange to the point of defiance.

In another photograph, a vast spider web of knotted shapes and patterns is stretched over laundry lines supported by two wooden poles (figure 2.2). Textile forms creep like vines, almost engulfing the frame of the photograph. In the lower right-hand corner of the photograph, a young man looks skyward. Is this web an avant-garde art installation, or a worn-out badminton net? To another observer, it might also recall the traditional raffia weavings from the Belgian Congo, the designs and materials of which had come to the attention of European anthropologists around the same time.[5]

This knotted creeping vine was camouflage material in its most effacive—and most iconic—so-called flat-top, form. Netting was the most prevalent form of strategic concealment employed during World War I and also the most effective. The young man in the photograph was standing inside a portrait of the new invisible fabric, literal and figurative, of warfare.

Even before the Special Works School opened its doors, experiments in using grayish-brown netting to obscure forms on the ground were taking place on the continent. Throughout 1916 and

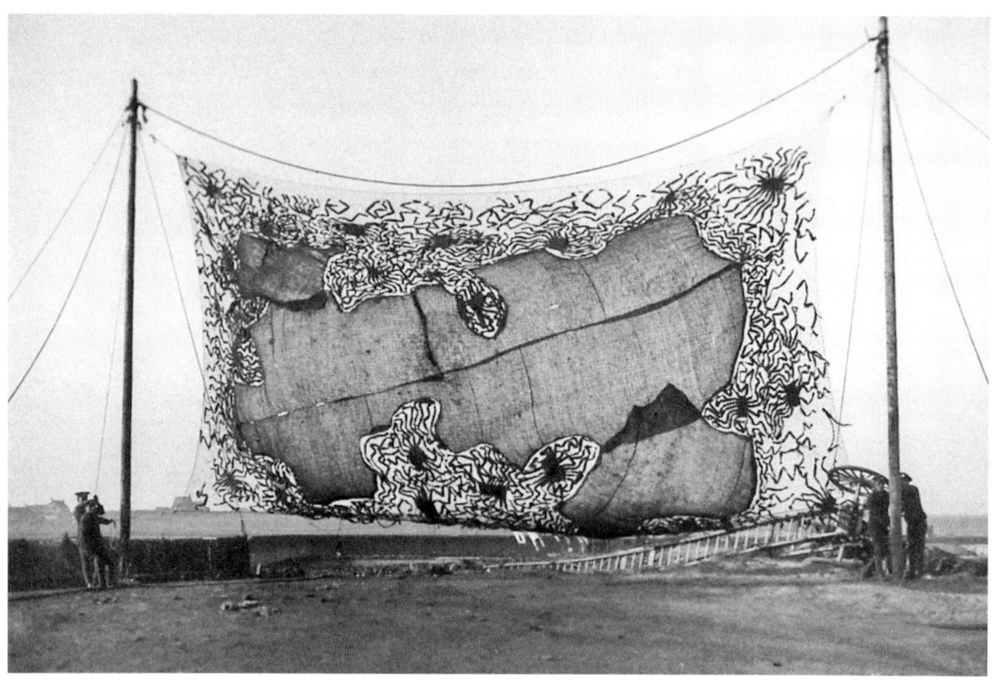

Figure 2.2—A new fabric of warfare. Netting is stretched between two poles, and garnished in a "flat-top" configuration near Arras, France, ca. 1917. Reprinted courtesy of the Imperial War Museum [Q1779].

1917, soldiers in France stretched nets across horizontal and vertical beams that had been erected over streets, houses, and cars (figure 2.3). Sometimes these nets of interwoven canvas and raffia, studded with clumps of painted hay, tree branches, or twisted rags, festooned entire villages. Homer Saint-Gaudens arrived in the French countryside with the Fortieth Engineers of the American Expeditionary Forces in 1918 and described a "magic veil" blanketing the scene. Netting spread like kudzu. According to architect and set designer Raymond Myerscough-Walker, it "was used to the extent of 7 million square yards during the war, and for various purposes."[6]

Viewed from above—from the perspective of airplanes overhead—the forms and patterns of the netting seemed to disappear. Instead only soft, rolling curves and shadows of pastures, fields, and woodlands appeared. Netting blotted out the villages, trenches, and artillery dumps; they became hidden parts, cloaked in secrecy. Clothing for the earth, netting was William James's overcoat writ large. But in contrast to Abbott Thayer's proposed scheme, this camouflage was not crafted in relation to a single "crucial moment" snapped out of time. Instead, it was made to hinder the derivation of meaning from change serially, measured over time. It was designed to stem the power of photography and to mar the confidence of photographic analysts to interpret what they saw. World War I's mixed-media netting thus served as a framework for both the environment and individuals. The net was once primarily a harvesting technology; now it was a framework—a flexible grid counterpoised by aerial photography's grid. Nets, when garnished and installed properly, made it hard, if not impossible to interpret a visual field effectively, thereby undermining photographic reconnaissance (figure 2.4).

From Certainty to Doubt

Until about 1914, people in Western Europe used nets primarily for the capture and containing of living beings. Mosquitoes could be deterred, fish or the occasional butterfly ensnared. Unraveling the evolution of netting as camouflage, however, is inextricably tangled with the technology-driven transformation of military intelligence.

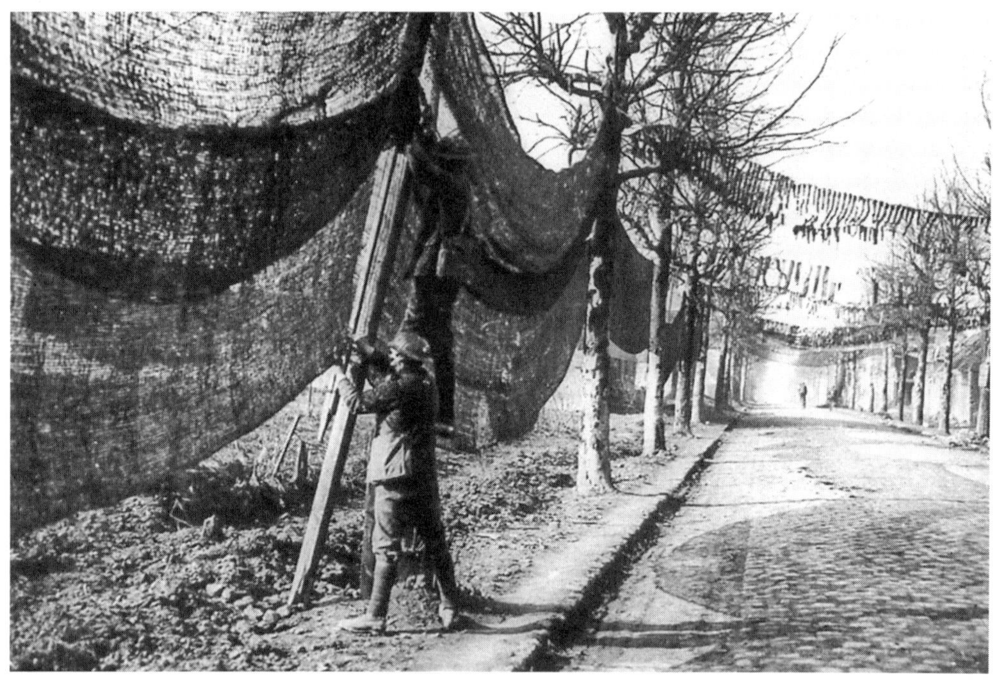

Figure 2.3—Streets under netting. Interwoven canvas and raffia, studded with clumps of painted hay, branches, and twisted rags, is arranged above and alongside the La Basse road near Cambrin, France, in March 1918. Reprinted courtesy of the Imperial War Museum [Q8553].

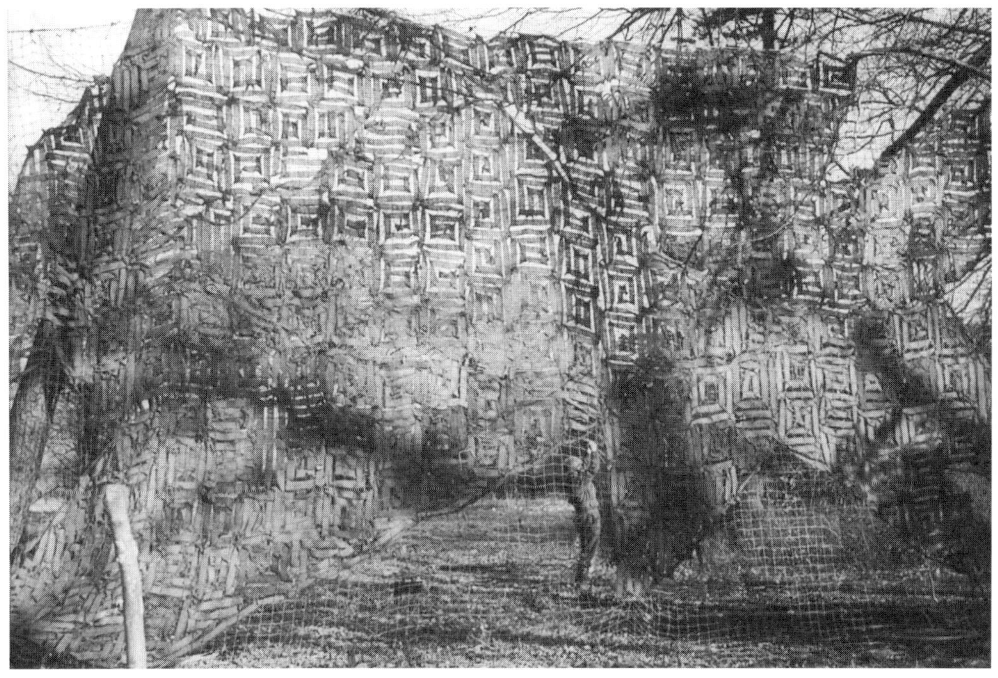

Figure 2.4—Netting as grid. Produced and garnished in great quantities, nets became a way to undermine the accuracy of photographic reconnaissance by eroding visual distinctions between subject and background, object and environment. Here a soldier looks up into the woven grid. Photograph by M. Puttnam, February 26, 1941.

Throughout the nineteenth and early twentieth centuries, military tacticians typically understood reconnaissance and surveillance as two separate and distinct enterprises. Systems of surveillance were developed to watch over suspicious parties. Prisoners of war or certain sectors of society, such as psychiatric patients or factory workers, were presumed guilty and placed under suspicion for deceitful activity; for sins real and imagined, they were watched.[7] Surveillance was employed to maintain control over these individuals and, if possible, to maintain their productivity also, such as in factories that enforced production quotas. These systems were predicated on the idea that the observers did not trust the observed. If left unchecked, the observed might shift position or appearance. These systems also assumed that the observers could remain unseen by those under surveillance.

Systems of reconnaissance, in contrast, generally referred to the less immediately fraught and more impersonal practice of surveying territory. In a military context, reconnaissance involved exploration on foot or horseback of the layout and resources of an area prior to its invasion.

The traditional object of reconnaissance was to see and note the world as it appears at first glance. To reconnoiter was to discover territory through personal, first-hand observation. In the event of reconnaissance in warfare, it was understood that you were surveying territory occupied by enemy forces. A dual faith grounded such missions from the Napoleonic era onward; faith in the model of human vision, on the one hand, and faith that the enemy had not engaged in systematic visual deception, on the other. The reconnoiterer, generally called the "scout," documented the visible characteristics of enemy territory at face value; for this he was granted both authority and accountability. Captains and scouts alike understood reconnaissance to be a singular (one-time) survey of an environment and its potential resources and natural defenses.[8]

The scout was held accountable in the sense that he was charged with coming up with a representative image that concluded his findings.[9] Noting down objects of interest as he encountered them, including dimensions and coordinates, he collated this information into a "field sketch" that constructed a viewer whose vantage point was that of a "commander on the hill."

The utility of the field sketch was bolstered by a system of visual icons designed to help the man who was coordinating the eventual military effort to identify objects of strategic interest. Training manuals on reconnaissance emphasized the need for scouts to keep their eyes open and implements ready. According to the 1884 manual of Captain Robert S. S. Baden-Powell, founder of the Scout movement: "The features, that is hills, roads, towns, woods, bridges, etc. are represented by certain signs that are called 'conventional' because they are agreed upon beforehand to indicate certain things. Thus all the reconnoiterers use the same signs and the officer in charge easily compares them."[10]

In the world herein constructed, things are precisely as they seem—which is to say, as their outward form implies. Meanwhile, it is clear which objects—"hills, roads, towns, woods, bridges, etc."— are worthy of identification and representation. The discursive field in which the manual was conceived and distributed is organized around a certainty that outward form does not deceive: what looks like a tree (referent) is actually a tree (entity in the world) and should be represented by a tree (sign). In other words, information, even when encoded and transcribed, should straightforwardly enable navigation and conquest.

The world to be investigated, whether from a balloon or on horseback, stood still as the scout moved through space, measuring implements and sketchbook in hand. This is not to say that the environment didn't contain hidden dangers, but these were the exception, rather than the rule. According to the *Scout's Handbook and Instructor* (1912), "the country is like an open book. Information is there in the form of signs. They only need to be read."[11]

By the start of World War I, however, the battlefield was no longer an open book. The development of the airplane, then simply called "the machine," precipitated a reconfiguration of warfare from the top down. Bombs could be dropped from the sky, and thus the field sketch and its view that privileged the "commander on the hill" was rendered wholly obsolete. Scouts took to the skies with their cameras, while photographic interpreters mined the photographs thus produced. *Camoufleurs* crafted ground cover designed to undermine this work. In the back-and-forth between these two groups and activities, new practices of hide-and-seek and hide again emerged. What looked like a tree was possibly not a tree at all, at least as far as reconnaissance was concerned (figure 2.5).[12] Rather, it was likely something to place under surveillance, something worthy of suspicion. Doubt subverted certainty that a thing must function as it appears to function.[13]

As concurrent advances in photography and aviation blurred the boundaries between surveillance and the old reconnaissance, netting production provided a means of mediating between the human, natural, and photographic realms. It was also both portable and changeable, potentially chameleonic. Netting became a visual and material medium of immersion in nature at the same time that it offered resistance to the photographic image (figure 2.6). By the end of the war, no longer could one assume that to measure and to notate properly what was observed was to know and transmit knowledge, at least where surveillance and the old forms of reconnaissance were concerned. In turn, photographic surveillance and mixed-media countersurveillance became the weft and the warp of modern reconnaissance.

Seeing and Interpreting

The first attempts at aerial reconnaissance were conducted without the aid of photography. In 1914, Brigadier General David Henderson, Royal Flying Corps commander and director of military training, began his new edition of *The Art of Reconnaissance* (first published in 1907) with this prefatory note about reconnaissance in the sky: "It must be expected that the principle dispositions and movements will be disclosed, and that the fog which formerly obscured the initial strategic design will be cleared away."[14] As Henderson continued, "the aerial scout can see what Wellington

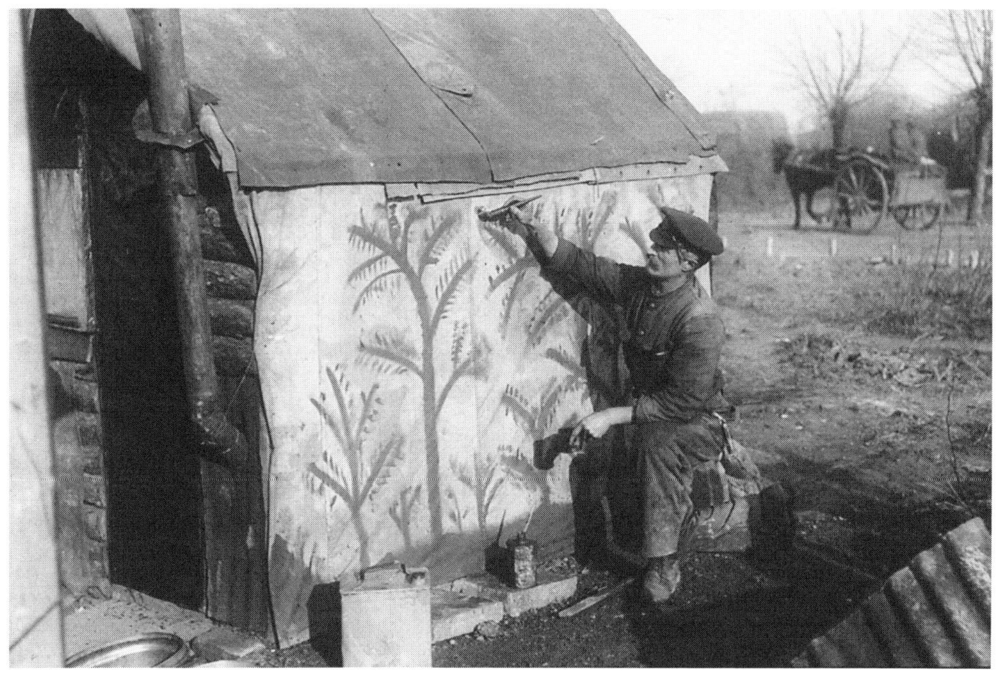

Figure 2.5—Tree graffiti. A member of the Army Service Corps paints trees on waterproof sheeting covering his living quarters near Boesinghe, Belgium, in late January 1918. Reprinted courtesy of the Imperial War Museum [Q10634].

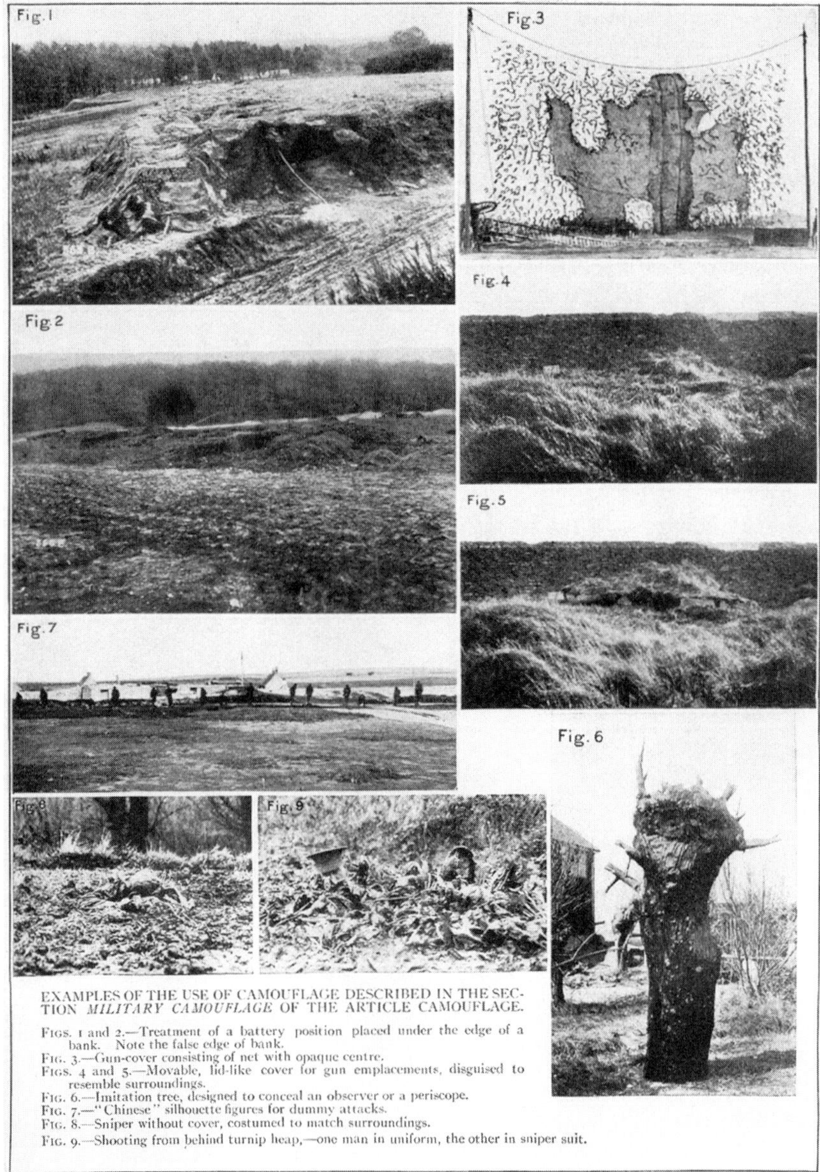

Figure 2.6—Camouflage defined. Lieutenant Colonel Francis Wyatt, a member of the Royal Engineers, explains the phenomenon s.v. "Military Camouflage" using net-based structures and effects as prime examples in the 1922 *Encyclopedia Britannica*.

always wanted to see, 'what is on the other side of the hill.'"[15] But the "fog" that troubled him was metaphorical, as well as literal. Aerial scouts still traversed the skies with only their notebooks and sketching implements in hand, drawing what they saw and what they remembered having seen. The camera, it was hoped, would make everything clearer.[16]

Still, as had been obvious since the first attempts at bird's-eye photography in the mid-nineteenth century, making aerial photographs presented unique obstacles to even the most intrepid photographer. In 1859, Félix Nadar struggled to produce what is thought to be the first aerial photograph by employing a cumbersome camera in a balloon tethered over the Bièvre Valley.[17] As early as 1913, RFC pilots made limited attempts at airplane photography, bringing handheld cameras into the air.[18] Photographing from an airplane presented many difficulties not present in balloon ascension, including the constant vibration of "the machine" and the added complication of working at altitudes of several thousand feet. As one World War I aerial photographer recalled, there was found to be "a veil of haze over the landscape when seen from high altitudes, and the consequent need for sensitive emulsions of considerable contrast, and for color-sensitive plates to be used with color filters. Aerial photography strains to the utmost the capacity of the photographic process."[19]

This wasn't all. Pistol cameras were cradled by hand and pointed out windows, producing exposures that, for all their detail, were neither perspectivally nor geographically precise. Camera shutters froze shut, and human fingers chapped and bled. Emulsion iced over, and exposed plates were ruined before development (figure 2.7).[20]

But the institutionalization of aerial photography had acquired an aura of inevitability by 1914.[21] In December 1914, the British Air Ministry established its first official Photographic Section.[22] This followed France's establishment of the first Allied photographic units weeks earlier. Technologies and strategic demands converged in the organizational practices of English—as well as French and American—photographic sections. Starting in 1915, the RFC stationed one photographic officer in each wing division. The photographic

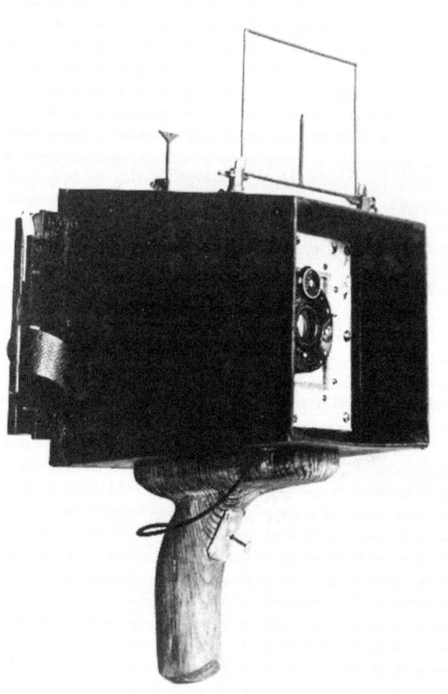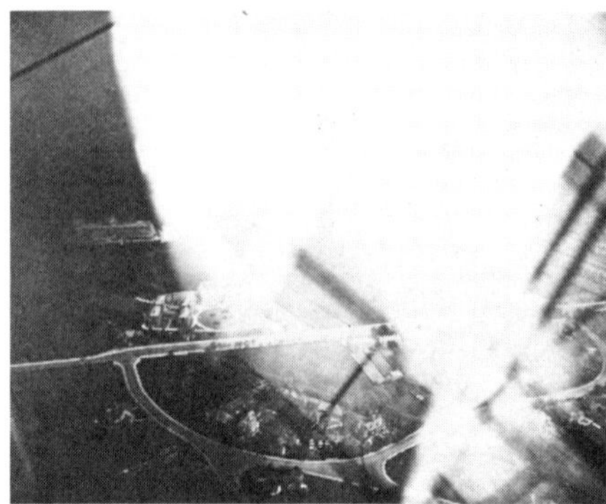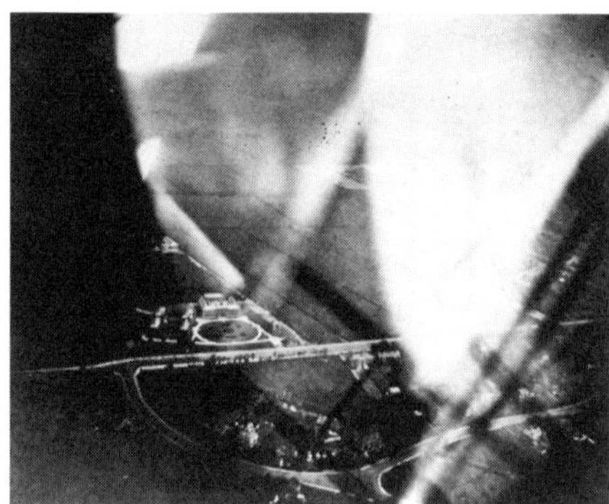

Figure 2.7—Pistol camera. Trial photographs made using this 1916 pistol camera show the ground glimpsed through wing and propeller. Reproduced from the May 31, 1916 "Report on Photography Section," Eighth Wing, commissioned by the Royal Flying Corps. Photographs made at Cramlington by Lt. Ellis. Communicated by G.S. Stoppard, Lt. Col. Commanding, Eighth Wing, Royal Flying Corps. [PRO AIR 1/128/15/40/175].

officer oversaw the technical work of photography in each squadron and also served as chief photographic interpreter. His work thereby combined activities of surveillance and countersurveillance.

Equipment upgrades drove much of the enthusiasm for photographic surveillance and reconnaissance. One key development was the propeller-driven camera with interchangeable lenses. Handheld cameras were replaced by fixed machine alternatives. The handheld pistol camera had prevented truly vertical photography, but by 1916, the most effective cameras were actually built into the airplane, tucked under the wing embedded in a wooden skeleton. This reduced some effects of vibration on the image and increased framing precision. Advances in emulsion and film technology also improved the quality of images, while at testing sites in England, photographic sections of the RFC and the Royal Naval Air Service (RNAS) developed techniques and technologies especially suited to high-altitude conditions.

By 1916, legible exposures could be made up to several thousand meters in the air. The L-type camera weighed thirty-seven pounds and had a motorized shutter. At fifty miles per hour in the air, plate change, powered by the propeller, took only ten seconds.

The decisive role of aerial photographs in determining British strategy in the Somme Campaign of 1916 transformed the status of photographic reconnaissance for intelligence purposes.[23] Field-sketching manuals gave way to textbooks on photographic interpretation. By 1917, the RFC and RNAS suggested that all pilots be trained in aerial photographic practice.[24] That same year, RFC photographic sections printed over four thousand aerial photographic negatives a day.[25] Mobile photographic laboratories printed many of these for immediate reconnaissance purposes—to visualize enemy territory and reproduce it in photographic form. Officers printed others to test the effectiveness of new concealment measures, such as those developed under the rubric of camouflage at locations such as the Special Works School in London.

In early 1918, the British War Office amalgamated the RFC and the RNAS into a new division, the Royal Air Force (RAF). The

RAF based photographic operations at the Central Photo Centre in London and, beginning a standardization of aerial photography technology, set glass plate size at four by five centimeters. The first Interallied Photographic Conference met in Paris later that year. At the 1918 conference, British, French, and American forces coordinated their photographic intelligence efforts; they further standardized technologies and scales of measurement.[26]

Once standardized as instruments of indirect observation, aerial photographs offered significant advantages over fly-by direct observation alone. Aerial observers working without cameras had to see things as they happened. The observer had time for notation, but not extended reflection and analysis as the airplane traversed the sky.[27] But the aerial photograph created a permanent visual record of a specific piece of territory at a given date and time. Airplane photography also enabled a geographical specificity lacking in older forms of bird's-eye photography. By 1915, engineers had figured out how to place a lens on the back of aerial cameras that recorded the direction of the compass needle directly onto the negative. Balloon photography had lacked this precision, a precision crucial for mapping, modeling, and analysis. With balloon photography, wind patterns or tether length or some combination of the two determined the viewfinder's purview. By contrast, aeronautical "machine" navigation made it possible to control exactly what was in and out of the frame.

Aerial photographic negatives and the prints they spawned became elements of a multivalent visual archive, one that commanded the attention of far more people. Seeing and interpreting, once simultaneous activities, became enterprises separated in both person and time. Long after exposures were made and film was developed, intelligence officers, commanders, *camoufleurs*, and photographers themselves pored over photographs with loupes and stereoscopic devices. They saw, indirectly and in good time, all there was to see.

Seriality and Concealment

The aerial photograph was a record, prime facie, and a mosaic of indexical signs of the real. Traces of light reflecting off surfaces bored themselves into photographic emulsion. Viewers gleaned grayscale details of shape, texture, and shadow from individual photographs. When individual photographs were arranged sequentially and compared, one had what photographer and writer Alan Sekula calls "evidence of systematic change."[28] A typical protocol could be stated as follows. Take two photographs from the same height and coordinates. Use the same aperture, shutter speed, and focal length. Make sure both exposures take place at the same time of day. Any differences between the two photographs could be read as potential carriers of meaning. Objects in a new position, lines in places where none had been discerned earlier—both spoke to what had happened in the temporal gap marked by the differences between the frames.

Through iterative applications of aerial photography, the repeated photographing of a site over time, it became possible to monitor minute changes in the landscape. The photographs were stacked up, scribbled on, and entered into an intelligence archive. One could compare and contrast results made weeks, months, even years apart. In some ways, these aerial photographs, once compiled into atlases, functioned within the context of what has been characterized as "mechanical objectivity."[29] As nineteenth-century critics of this so-called objectivity sometimes pointed out, at times it could be hard for users to locate the subject of the individual photograph. Even if the "where" were uncertain, however, the "what" was clear, and remained unquestioned. Neither photographers nor interpreters questioned the motives of the "what." Willful deceit on the part of the photographic subject (whether geological formation or indigenous tribesman) was virtually unthinkable.

And yet in contrast, the aerial reconnaissance photograph—with all of its privileging of mechanical exactitude—is rooted in a fundamental uncertainty about the identity of the "what." Subject and background intermingled in a confusing visual field.

Aerial photographs depart compositionally from many other types of instrumental photographs, whether static single or serial images. Like Thayer's feather paintings and habitat photo collages, serial reconnaissance photographs, even though they represent three-dimensional scenes, lack a fundamental differentiation between foreground and background. Three dimensions register as a flattened abstraction, and thus, not surprisingly, early aerial photographs bear a compositional resemblance to modernist photography. Art historians and scholars of modernism have linked developments in cubism and abstraction—particularly the work of Bauhaus artists such as László Moholy-Nagy and Alexander Rodchenko—to the proliferation of aerial photography and the bird's-eye view.[30]

Even setting camouflage aside, aerial photographs were difficult to "read." Learning to interpret aerial photographs was a process of instruction in the terms and conditions of visual diagnosis (figure 2.8). Visual training, so called, taught soldiers how to see critically and skeptically and thereby be better able to arrive at information reliable enough to inform strategic military decisions and save lives.[31] Rather than celebrate the fracturing of foreground and background, as abstract art photography aspired to do, the photographic interpreters aimed to resolve the confusion.

Military photography was strongly influenced by earlier uses of serial photography in the disciplines of physiology, geography, and industrial psychology. In particular, reconnaissance photography—both its production and interpretation—drew on the tradition of late nineteenth-century chronophotography, the analysis of movement through the sequential production of photographic images. In the 1870s and 1880s, the physiologists Etienne-Jules Marey and Eadweard Muybridge, working in France and in the United States, respectively, seized upon the potential of instantaneous photography and applied it to the study of the locomotion of various bodies through space. Chronophotographic analysis was serial photography at its core.

Eadweard Muybridge, for example, created long series of instantaneous photographs within the confines of his specially constructed outdoor studio. As Thayer would later turn the Harvard Museum of Comparative Zoology courtyard into a stage, Muybridge turned an athletic facility into a laboratory. By the mid-1880s, Muybridge was using a battery of sixteen electrically synchronized cameras, each outfitted with small glass negatives. Muybridge spaced these cameras

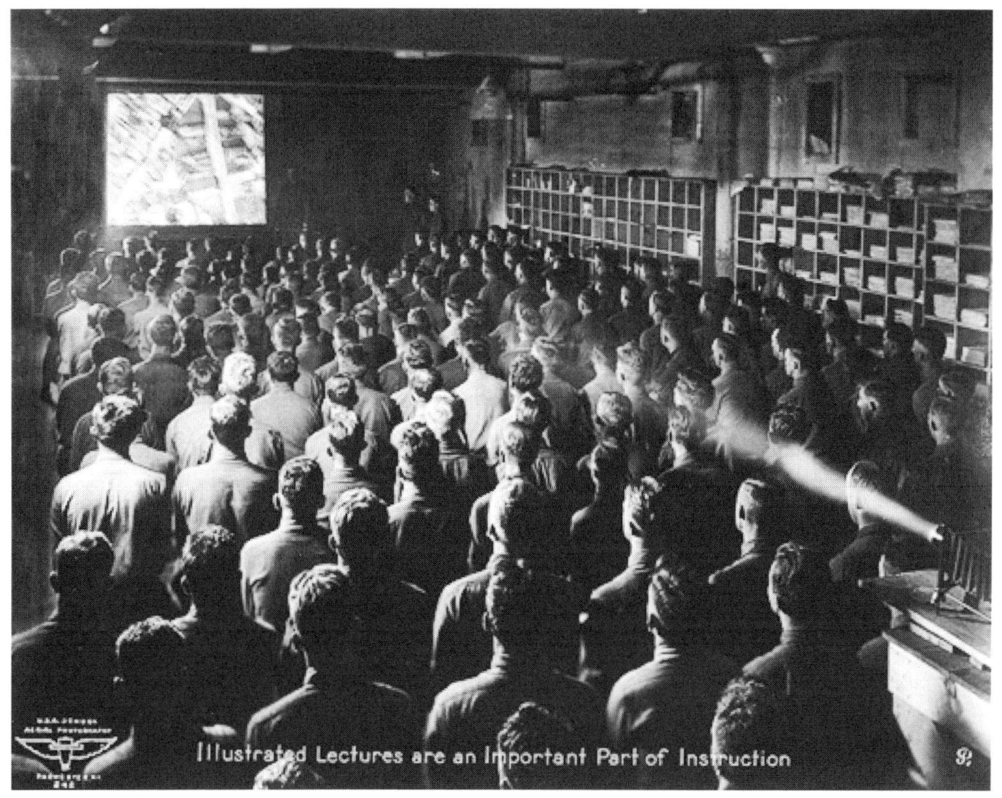

Figure 2.8—Visual training. In this 1918 documentation of an aerial photography training session, intelligence officers from the School of Aerial Photography in Rochester, NY learn to interpret aerial photographs. Reprinted courtesy of the George Eastman House, International Museum of Photography and Film.

evenly—serially—along the length of the run in his quest to dissect human and animal motions of all kinds.

After developing the semitransparent positives, Muybridge labeled them and placed them in sequence on a glass plate. A gelatin contact print served as archival negative for reproduction.[32] The progression of measurable changes between one image and the next revealed the dynamics of movement heretofore mysterious. The chronophotographer channeled movements into a comprehensive archive of motion, creating a narrative that unfolded from left to right, from three perspectives.[33] The individual frames each constitute an attempt simultaneously to mobilize and to control spontaneous motion.[34]

In this mode of serial photographic production, ideally, all variables but the one in question could be controlled. The subject moved as the environment (in this case, the laboratory background) stayed still. The rigid compositional elements of the resulting images effectively highlight their maker's attempt to control organic movement. The absence of spatial extension and hue gradation heightens the sense of nature rendered orderly; three dimensions are flattened into two-dimensional docility and legibility.

Everything moves from left to right; capturing the essence of bodies in motion across time thus comes to seem as simple as reading words on a page. Solid black borders simultaneously frame and divide the rectangular blocks from one another. The bars' symmetrical gridding mark out a strict compositional skeleton, and each frame is demarcated by a precise moment and location of exposure. Horizontal stripes—bands of white (background), gray (ground), and black (borders)—bind together the rows of frames.

Photographic production and interpretation of this kind required such a fixed background. The subject should move and the environment should stay still. Muybridge indeed aspired to an experimental system in which all parameters except a single unknown could be manipulated. In theory, nothing was hidden from view.

The practice of World War I serial photographers required a different, but related set of assumptions and apparatus. The military field to be photographed, unlike Muybridge's studio, was a laboratory of the unknown. These spaces had no clear lines demarcating foreground and background. Reconnaissance aimed to decipher change in a man-made environment over which the analyst himself had no control. The goal was to determine the "facts" of the enemy and to distinguish precisely the kinds of foreground/background relationships that Muybridge made sure were fixed well before the documentation process began. Determining the lines of change from one frame to the next was a process of discovering precisely those elements of the photograph that were of interest. This was not easily done.

Officers compared photographs made over the course of a single day, sometimes hour by hour, to get a sense of typical diurnal lighting conditions. Series produced from the act of photographing a given site monthly or seasonally provided a baseline of the natural environment as it progressed through time. Aerial reconnaissance photographers made repeated images of a particular site from a given altitude and angle. The mission returned daily, weekly, or monthly to monitor potential man-made changes to the territory below.

Intelligence officers pored over photographs coming in from the battlefield to help decipher critical information about possible changes in position, traffic, and enemy construction sites. Squadron Two of the RNAS, in collaboration with intelligence officers at Dunkirk, for instance, produced a report including this diagram based on a body of aerial photographs, documenting changes over time at an enemy artillery battery (figure 2.9). The report noted: "Photos were taken with the object of seeing the effect of the recent fire by the French artillery on Tirpitz Battery. A comparison of the current photographs with 790 of 24/4/17 shows nine new large shell holes.... Between 759 of 13/4/17 and 790 of 24/4/17 the most part of the small black squares on the face of the battery were removed but have since been replaced (*vide* present photos)."[35]

The negatives—both the "current" ones and those of "24/4/17"— had been made at the same time of day, from the same perspective, and under comparable atmospheric conditions. In the geography

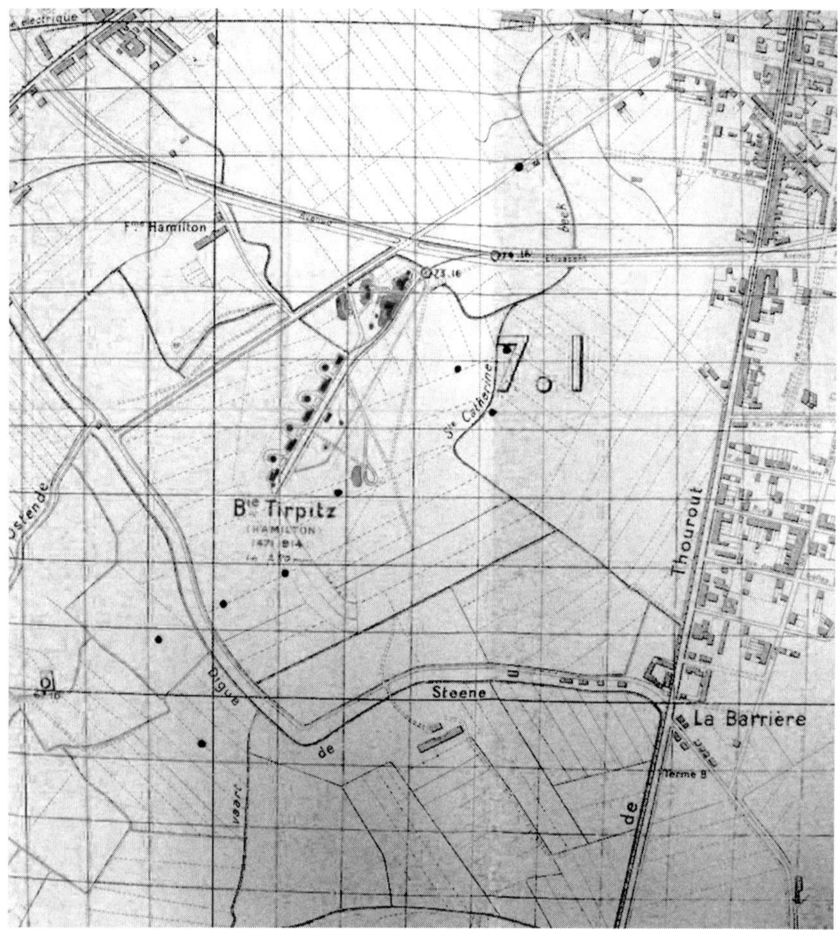

Figure 2.9—Tirpitz battery. Diagram based on a study of a series of Royal Naval Air Service aerial photographs, produced to track changes in enemy positions, camouflage, and construction sites. [R.N.A.S. Studies no. 1 to 50, April–July 1917, AIR 1/81/15/9/201].

of the unknown, photographers aimed to control and quantify as much as possible—to simulate, or at least approximate, laboratory conditions. The chief author of the report indicated suspected alterations in the landscape, purposefully concealed or otherwise, in the attached diagnostic sketch.

In the act of interpretation, signals had to be separated from noise in raw images cluttered with shape, shadows, and patterns of all kinds. As Sekula determines, "With the development of camouflage…the indexical status of the sign was thrown into question, thereby inflating the suspicions of the photographic interpreter."[36]

To aid in the process of reading the hidden signals, aerial photographers and their cameras approached the field from different angles. They made photographs from both vertical and oblique perspectives. Comparison of a series of oblique and vertical photographs and their synthesis through aerial stereopsis enhanced perception of depth and relief from an aerial perspective.[37] Stereography—stereographic viewing and photography—facilitated photographic interpretation at this level. Stereoscopy exaggerated perspective, making conditions of volume more clear than would otherwise be the case.

RAF officers also laid out photographs taken at regular geographic intervals so as to overlap partially (figure 2.10). With these collages, huge swaths of enemy territory could be reproduced; examination of the points of overlap with stereoscopes then provided insight into the concealment strategies being deployed by the enemy. Suspected enemy activities came into view. Intelligence Headquarters issued reports, dispatching photographic prints and diagnostic sketches to air, camouflage, and ground officers. The visual signs on the sketches signified activities, rather than static objects. These signs marked points of change, pointing to the fluidity, rather than the fixity of boundaries and structures in that space.

Interpreters began to detect decoys (traps), mimics (distractions), and attempts at the large-scale concealment of the movement of troops, trench lines, or supplies as *camoufleurs* fought back against the surveyors from above. Over time, *camoufleurs* developed astute

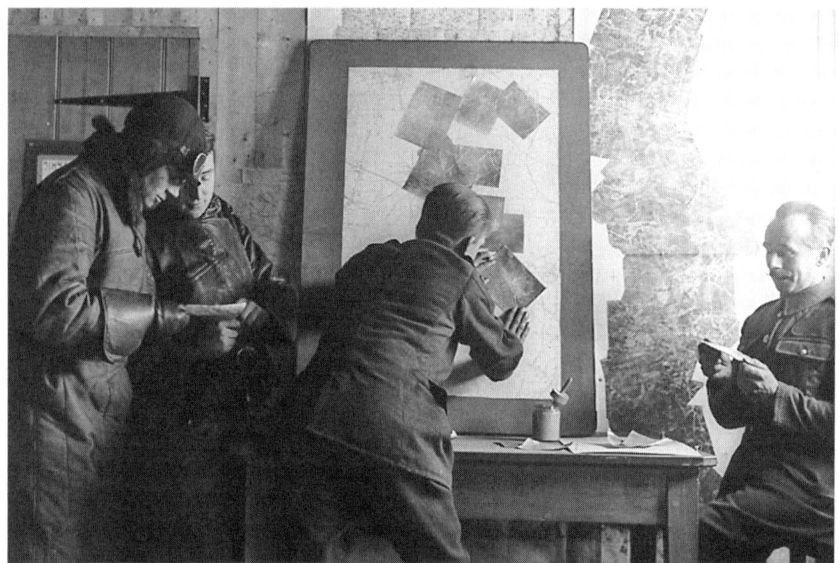

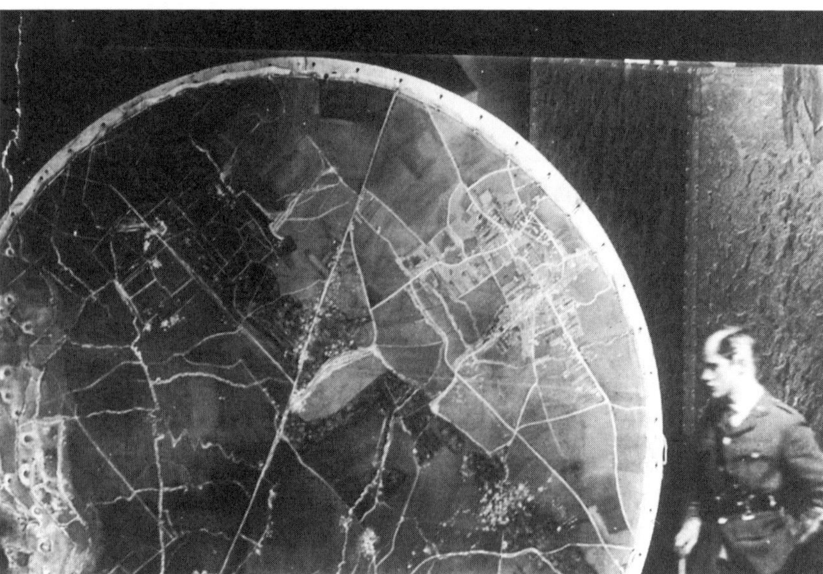

Figure 2.10—Collage and mosaic as military tactic. Top: Photographs taken over enemy territory are plotted against a map at an Royal Flying Corps office near Arras, France, February 1918. Bottom: Richard Carline, official war artist, poses next to a painted aerial panorama constructed based on such photographs, February 1918. Photograph of Richard Carline by Sidney Carline, Carline Collection, reprinted courtesy of the Imperial War Museum [Q8533].

means of making photographic details smudge together. The lines that rendered objects as discrete visual forms in a reconnaissance photograph might be massaged away. And if the *camoufleur* were able to erase the visual traces demarcating one object from another, reading "between the lines" would be no easy task. In the words of British artist and officer Solomon J. Solomon (1860–1927), "what is not seen is for all practical purposes non-existent."[38]

"The Material"

Over the course of days, months, and years, camouflage netting infiltrated the landscape itself, as well as the framed contents of reconnaissance's photographic units. As the 1916 Intelligence Report produced by Squadron 2 of the RNAS noted: "Alterations have been made in the patches of camouflage both on the north face of the emplacements and also along the southern side of each side of the feeding railway. It is not considered necessary to re-plot this camouflage on each occasion that the battery is photographed, but it is interesting to note that since 529 of 10/11/16, all subsequent photos of the battery show this protective disguise."[39]

"Patches of camouflage" in the reconnaissance photograph indicated a job badly done. The reason it was not necessary to "re-plot this camouflage" is because it had already been found. Conspicuous camouflage was worse than no camouflage at all; it made a mockery of aspirations to strategic concealment through protective coloration. This did not mean that camouflage had to be modified day-to-day per se. It just had to blend in enough on any particular day not to be noticed. Inconsistencies revealed (bad) camouflage (figure 2.11).

Early modes of camouflage—such as disruptive painting and the construction of dummies and decoys—were unwieldy, labor intensive, and often wholly or largely ineffective. Solid dummy vehicles and permanently painted weapons were hard to transport; they also often aroused suspicion. Whereas layers of netting could be endlessly reconfigured, coats of paint dried solid; they were permanent, rather than easily changeable. Dummy silhouettes and wooden heads mounted on sticks, as well as disruptive paint schemes for guns,

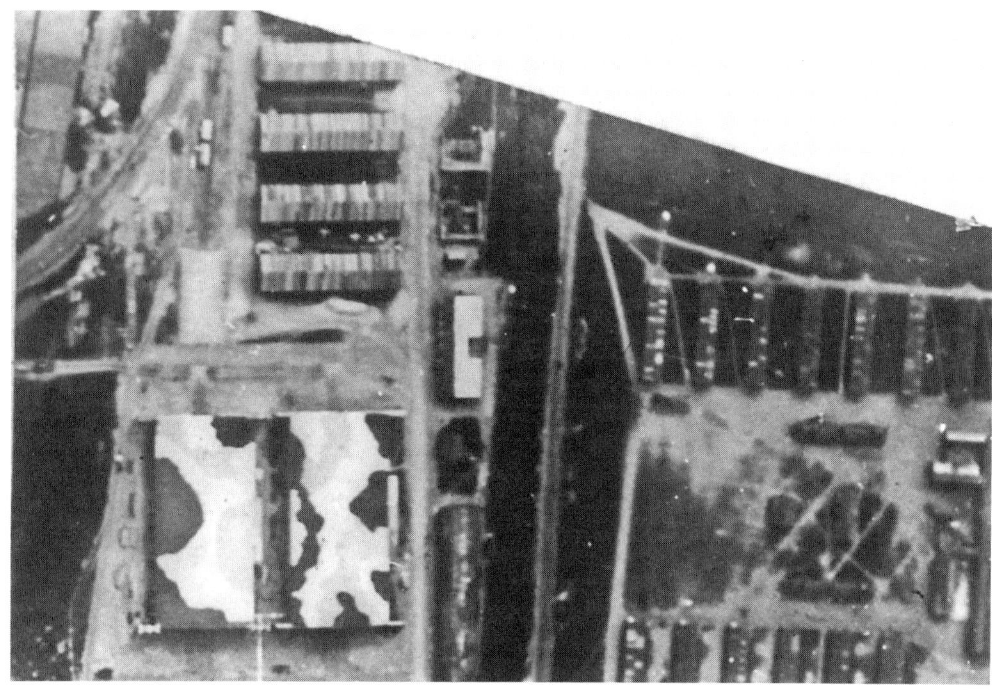

Figure 2.11—Bad camouflage. Camouflage done poorly proved easy to spot, as in the lower left corner of this World War I instructional photograph. Reprinted courtesy of the Imperial War Museum [Q17744].

howitzers, and tanks, were piecemeal deception of the human eye or sniper scope on the ground. Incomprehensive, they rarely fooled the camera. Also, vehicle transport radically decreased the effectiveness of painted camouflage. Once a tank had been painted to suit one position, which is to say, "to look like something else," it was difficult to recamouflage it to suit another position or even to get to that new position without detection.

To make good camouflage—which is to say, the kind of "patches" that wouldn't show up as such, as they do in figure 2.9—one had to understand how photo reconnaissance worked and also project oneself mentally into the aerial camera position. Accessing or at least envisioning the viewpoint from which trenches and traffic flows should be hidden provided insight into appropriate material measures. This was camouflage in a nutshell as determined by the regime of serial aerial photography.

Artist-turned-*camoufleur* Solomon J. Solomon claimed to have invented, or rather to have innovated, camouflage netting, aka "the material." A British oil painter known by the early twentieth century for his society portraits and mythological scenes, as well as for his prominent position within the London Jewish community, Solomon's origin story for "the material" begins with a moment of epiphany one evening in August 1914. He sat in the backyard garden of his mother-in-law's vacation home in St. Albans, just north of London. He fondled a scrap of butter muslin that had made its way outside, like a piece of canvas gone astray. Solomon had just been reading about enemy aerial bombing and observation methods. Even at this early date in the war, which had broken out earlier that summer, German pilots were using photography to ascertain the effectiveness of bombing efforts.

With this in mind, Solomon held the semitranslucent, flexible cloth up to the light; it hovered several inches above ground, over the saplings and flowerbeds. He examined the effect.

As the floral details disappeared from his view, an idea coalesced. One might drape objects in such fabric, dress them so that "they should not be seen, in the hope that it might be the means of saving some lives in the war."[40] Solomon obtained more muslin strips and some fresh stalks of bamboo. He painted canvas rags in shades of green, brown and gray, then applied patches of moss and dead leaves. In his backyard workshop, he bound rags and leaves into bundles, which he knotted to the muslin. Using the bamboo stalks as impromptu tent poles, he stretched the decorated fabric first over the saplings and then over miniature "trench lines" he had dug between rows of flowers.

Solomon's experimentation in outdoor obliterative countershading, he felt, was guided by his experience as volumetric painter and art instructor.[41] Similar experience had guided Thayer's study of protective coloration in the animal world. Unlike Thayer, however, who had worked on individual organisms modeled in skin, clay, or emulsion, Solomon worked on the earth itself. As a trained oil painter and instructor, Solomon had long told his students that "it is by variations of light and shadow, often very delicate, that one recognizes that an object is a solid and is detached from its background." Solomon continued this thought in his textbook on oil painting: "By reversing nature's lighting and by eliminating shadow, flatness results."[42] Solomon translated these ideas into camouflage principles, as in his *Visual Deception in Warfare*, wherein he concluded that "the elimination of shadow is the essence of invisibility."[43]

This conclusion was not immediately arrived at. The "observation post tree," or "O.P. tree" for short, was one of the first major productions of the camouflage factories in 1915 and 1916.[44] The goal was to craft a mimetic representation of a tree—and not just any tree, but a particular tree at a specific site (figure 2.12).

To develop the O.P. tree, Royal Engineers representatives selected, measured, and photographed the original tree, in situ, extensively. The ideal tree was dead; often it was bomb blasted. The photographs and sketches were brought back to the workshop, where artists constructed an artificial tree of hollow steel cylinders, but containing an internal scaffolding for reinforcement, to allow a sniper or observer to ascend within the structure. Then, under the cover of night, the team cut down the authentic tree and

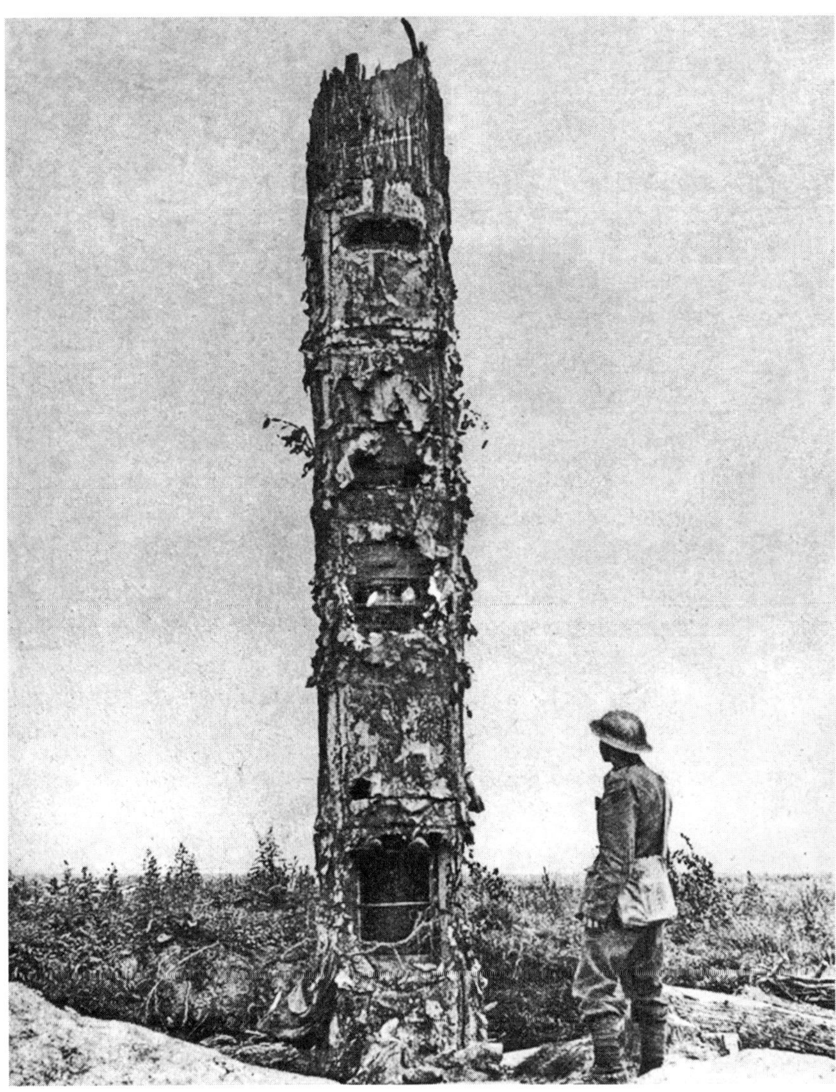

Figure 2.12—Observation post tree. Observation post trees, such as this one, allowed soldiers engaged in reconnaissance to report on their surroundings unobserved. Original caption (translated): "The English soldier located inside the tree shows the observer's position." Photograph and article printed in French newspaper *Le Miroir*, February 10, 1918, p. 13.

dug a hole in the place of its roots, in which they placed the O.P. tree.[45] When the sun rose over the field, what looked like a tree was a tree no longer; rather, it was an exquisitely crafted hunting blind, maximizing personal concealment and observational capacity simultaneously.[46]

Back in St. Albans, however, objects and the boundaries between them had been rendered hazy. As in an impressionist painting, demarcations between figure and ground receded into the grass before Solomon's eyes, and herein lay greater potential. Solomon wrote up his ideas and implementation strategy and submitted them to the British and French War Offices for consideration. According to Solomon's account of his presentation to the British Army at the Woolwich Dockyards, "the officers present were enthusiastic," yet he received no credit for the proposal from the War Office. Likewise, when he sent screens and instructional diagrams to French officials—then busy applying paint directly to heavy artillery—his proposal was turned down. Undeterred by rejection and fiercely patriotic, Solomon joined the military effort nevertheless. In the months following the demonstration, he began organizing the first British Camouflage Section while authoring pamphlets on practices of visual deception.[47]

By 1916, nets and screens similar to those Solomon had demonstrated at Woolwich littered the Western Front. Gun covers and trench covers became increasingly common in the workshop and the field. An unresolved question is whether or not Solomon's idea was stolen, artfully adapted, or redundant upon introduction.[48] Reflecting back on the development of "the material" from a distance of thirty-five years, Lieutenant Colonel Clement H. R. Chesney, commanding officer of the Special Works Factory in Arras, France, reflected "who can claim to be the inventor of these [nets] it is impossible to say."[49] According to camouflage officer Adrian Cornwell-Clyne (né Adrian Klein), later to become an expert in color cinematography: "'Camouflage material' means one of the following: scrim, fish netting threaded with painted grass, mesh wire netting, wire netting threaded with grass, wire netting threaded with canvas strips."[50]

Not quite invented, netting evolved, one might say, an adaptive and multipronged response to the new conditions of photographic warfare. Lieutenant Colonel Chesney reflected that the nets, "like all new inventions," had emerged as "the result of a number of people working at an idea." But as he reminded former officers in his *The Art of Camouflage*, "it is not so much the idea that is the invention as the translation of it into practical usefulness."[51] Solomon should be seen, in fact, as one among many realist artist-engineers for whom the idea of textile subterfuge and its application converged in material innovation on the ground.

Camouflage as Net Work

In early March 1916, the British War Office organized a camouflage unit in a disused factory in northern France in the seaside town of Wimereux. Solomon, initially a private with the Artists' Rifles civilian corps, but then granted the temporary rank of Lieutenant Colonel, oversaw the Special Works Park Wimereux, established by the Royal Engineering Force (REF), for the first six months of its operation.[52] Soon after his arrival, an engineer came in—Lieutenant Colonel Francis Wyatt, a member of the Royal Engineers—to supervise operations. Wyatt and Solomon allegedly did not hit it off so well; after six months, Solomon received orders summoning him back to London. His time on the front was over, and he returned home to become chief artistic advisor at the Special Works School in Kensington Gardens commemorated in Mottershead's album. Wyatt reported in his war diary that early goals for those employed at Wimereux were "to cover trenches, guns, etc., that required to be concealed from enemy observation by materials...treated in such a way as to make them merge into their natural surroundings."[53]

Wyatt's camouflage section aimed to supplement and enhance the capabilities of the French camouflage sections, which had been in operation since early 1915. He brought officers from England to set up satellite operations along the Western Front in Aire and Arras. The factories brought together British artists, military engineers, and French women evacuated from neighboring regions that had sent their men to the front.

Initially, the camouflage factories, especially the ones set up early on by the French Camouflage Section, worked on constructing observation posts made to look like horse corpses and blasted tree stumps in order to escape horizontal detection. Solomon's first assignment at the Special Works School was to oversee the design and construction of observation post trees—this in early 1917. By this time, periscopes, sniper scopes, and sighting devices had enhanced human vision to distances of up to sixteen hundred yards. Indeed, when the French and British began systematic work on camouflage, most of their work was restricted to cardboard and paint, and success arrived first along the horizontal axis. Using the media of painting and set design, camouflage researchers developed dummy silhouettes and disruptive paint schemes for tanks and artillery.

But because vertical detection via aerial photography—not to mention the increased accuracy of artillery—posed the most pressing threat of the time, they soon shifted focus.[54] As *camoufleurs* began to enlist photography to test their work, it became clear that painted camouflage was often ineffective. What was the most economical, effective, and portable means of confusing the enemy's photographic interpreters? A method for confusing the boundaries of objects and buildings in a way that would not be detectable in patterns of shadow, texture, or hue—the three qualities that always show up in photographs—was the answer. The photographic plate should register the smooth appearance of a rolling field. As Wyatt noted in 1922: "The material of which the camouflage is composed must at all times appear on the photograph like the object or surface it represents, and likewise appear natural to the observer's eye."[55] Military officials soon recognized the relative effectiveness of netting and textile-based vertical concealment measures such as those Solomon had envisioned that afternoon in St. Albans.[56] Camouflage should form a smudge that erased itself.

Meanwhile, an issue that developed in Solomon's mind was the question of what the Germans were up to. Countersurveillance involved both hiding from enemy surveillance and figuring out where the enemy was hidden. Amassing information about the German reconnaissance machinery was naturally of great interest. Such information would not only help the Allies improve their camouflage measures, but also help improve their aerial photography, because German optics were known to be far superior to their British and French counterparts. The Germans weren't about to give away their camera tricks, however. The most viable approach was harvesting the wreckage of downed planes and digging in the pockets of dead men. Occasionally, camera operation manuals were found in the possession of pilots or infantry killed in action. Mangled cameras picked out of heaps of exploded airplane bits provided important clues as to enemy technologies in current use. A kind of "camera autopsy" might be performed; the analyst assembled all the discovered pieces and tried to figure out how they might have fit together.

The related question, of course, was what the enemy's camouflage looks like, an especially hard one since effective camouflage would never be seen at all. The art of the *camoufleur* thus easily lends itself to extreme paranoia. Once you have camouflage on your mind, you are likely to see it everywhere. Solomon became convinced, for example, that the Germans had camouflaged everything; Germany increasingly became imagined, in his mind at least, as one giant piece of butter muslin. Indeed, after the end of the war, Solomon published a book called *Strategic Camouflage* in which he infamously contended that the German system of camouflage had been virtually absolute, although it had left no traces whatsoever, and hence the claim could not be disproved.[57]

Solomon's paranoid grandiosity notwithstanding, *camoufleurs* on both sides of the war *had* become adept at crafting the visual illusion of absence through degrees of mimicry. Parts of the net that would be most directly in the sunlight were more heavily laden with foliage and brush. The bundled rags in these parts should be painted a darker shade of green or brown. Meanwhile, on the outskirts, where the netting sloped toward the ground, the patterning needed to be substantially lighter in hue. The specifics of decoration were contingent on the topography of a particular piece of ground. *Camoufleurs* and camouflage officers internalized Thayer's law, whether know-ingly or not. And by the time the United States entered into the war, reporters were referring to Thayer's law of protective coloration in relation to camouflage practices at American training military camps.

And as Allied *camoufleurs* became better at conceiving how to assemble effective camouflage netting, they also became more effi-cient at producing it for the war effort. By 1916, the factories at Wimereux, Aire, and Arras had adopted the methods of mass pro-duction. Dried raffia palm imported from the French colony of Madagascar was dyed in a large warehouse. Reams of canvas were rolled onto the fields where they were sprayed with a mixture of paint and brown, gray, and green fabric dyes. The raffia and canvas were dried, tested for colorfastness, and then brought into adjoining makeshift warehouses to be cut into thin strips, one inch wide and twelve inches long. These strips were knotted by fishermen into net-ting consisting of one-inch squares and cut into sizes ranging from twenty to fifty square feet. Green, tan, and brown bundles were packaged for distribution to the front, where individuals applied raf-fia and local foliage as the environment required.

Meanwhile, large sheets of canvas were painted brown and green—and left intact. These canvas sheets, generally about twenty-five by twenty-five feet, provided a backing material and supplement to the smaller nets. In the field, a painted canvas sheet might be used to cover an ammunition dump. A gun that might poke out from the canvas would likely have been shielded with raffia-decorated netting, though supplies of the raffia ran low as demand exceeded imports. Over time, canvas replaced raffia, even for purposes of garnish. Canvas strips were found to be more durable; their color did not fade, and the texture did not degrade as quickly.

Careful deliberation over which raw materials to employ was essential. Chesney reflected: "Camouflage in practice is not so much a matter of making dream pictures as of organization and practical adaptation of materials.... As to material it must be light, strong, impervious to weather, fireproof and easily manufactured."[58]

Natural materials lacked endurance. The very thing that made them ideal—their availability in nature—also made them transient. Oil painted onto canvas, although perfectly adapted for portrait painting, also had several drawbacks when it came to camouflage material. Linseed oil in the paints broke down, and when placed near a heat source, the painted canvases would often burst into flames. This was a problem in the workshop and an even greater problem on the battlefield, where it caused several fires. A Belgian camouflage facility almost burned to the ground. Glue-based paints were less flammable, but worked only on vegetable fibers.

While material scarcity forced innovation and evolution in camouflage production, so too did the diversity of the people involved. Initially, the camouflage factories were largely staffed by enlisted British artists and engineers. As demand surpassed production, chief officers at the Wimereux, Amiens, and Aire factories began seeking out French female refugees.[59] Meanwhile, an agreement between the French and Chinese governments allowed an influx of temporary laborers from China to obtain work as painters in the camouflage factories.[60] By late 1917, French women, Chinese men, and British officers worked together in a large shed preparing nets, stitching up canvases, dyeing raffia, and drying painted canvases. The bottom image in figure 2.13 shows Chinese men applying paint through the spray pumps under the watchful eye of British commissioned officers, in a doubling of the surveillance thematic in which their larger activities were inscribed. In a corner of the other image a shed interior, the French women garnished nets as often as time and supplies would allow.

Collaboration between the artists and engineers was alternately productive and contentious. Artists tended to criticize the engineers' handling of paint and materials. Many claimed that "artistic vision" was needed for the proper application of contour lines and textures. Meanwhile, engineers accused the artists of impracticality. "It is quite easy to create camouflage in the brain, any imaginative person can achieve that, and it is quite another thing to reproduce the picture in material objects," Chesney the engineer reflected.[61]

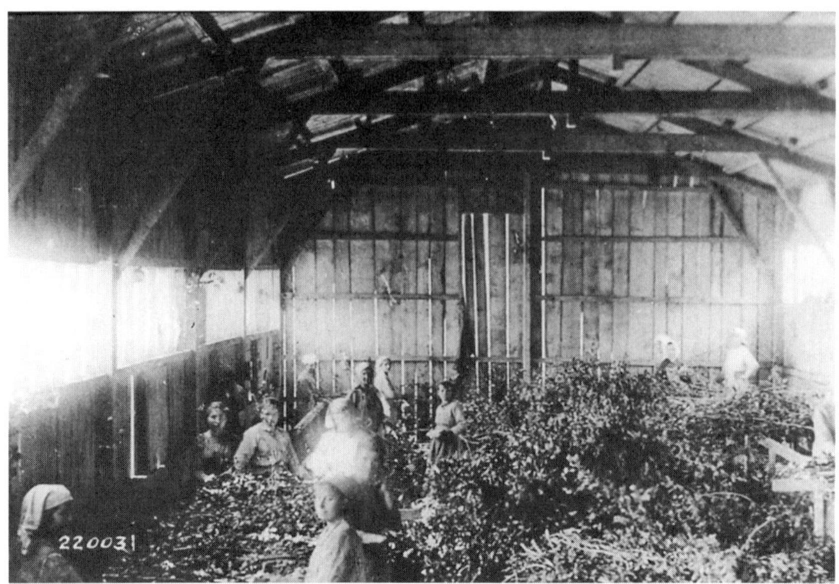

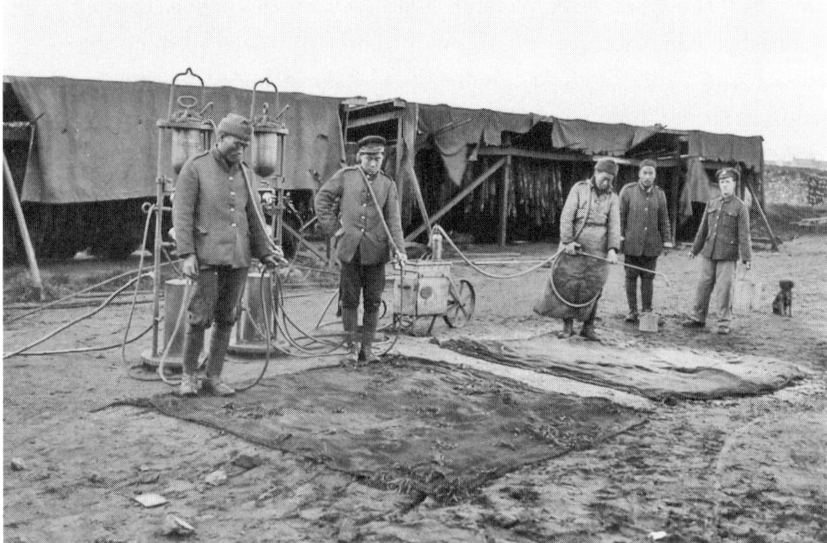

Figure 2.13—Camouflage workers. Top: French women at a camouflage factory, ca. 1917. Bottom: Asian camouflage workers use spray paint to cover nets laid on the ground near Arras, France. Reprinted courtesy of the Imperial War Museum [Q17801 and Q17815].

Although studio artists in France and England had developed the first preliminary camouflage models, by 1916, engineers were overseeing the camouflage factories throughout France.

From the studio, to the factory, to the field, net creation grew in complexity. A roller paint system was developed at Wimereux. The drafted Chinese laborers were soon pulling the raw canvas between two metal cylinders. Various cutting machines were employed to slice the painted swaths into pieces. These pieces were then used to garnish fifty, sixty, sometimes seventy nets a day. The women, who brought to this job skills in crochet and embroidery, among other textile art forms, knotted the fabric into flexible "textile grids" into which pieces of raffia were tied.

Individual female workers were given the task of "dressing" the nets, often one by one. At the Amiens workshop, the net was hung over a beam and its dresser walked around it with four baskets—each containing strips of brown or green canvas or scrim, two shades of each color. When the army ran out of green paint, yellow paint was used. Painters, themselves under the watchful eye of the engineers in charge, oversaw the women, who were dressed in smocks supplied by the quartermaster. Officers and artists in charge would sometimes provide the women with schematic drawings, at other points providing written instructions, such as "this net is intended for use as a, b or c." It was found that some objects showed through the scrim; in these cases, nets had to be laid beneath nets.

The nets were not complete when they left the factory, however. Soldiers in the field, once they received an assortment of nets, canvas, and raffia, had to develop their own site-specific designs for concealment.[62] The netting reached the soldiers on the line at regular intervals, along with food and artillery supplies. As soldiers received the netting, they gathered to garnish and modify it in relation to the particular environment in which they found themselves. Of course, they themselves could not literally see the environment from an aerial or photographic perspective, so they had to extrapolate—insofar as possible—from what they saw around them on the ground to what their environment would look like from the air.

Training and advice from a dedicated camouflage officer was of great service in this regard.

Figure 2.14 shows British soldiers on the line, tying local foliage into the netting they had been sent from the Special Works Park Headquarters. Here, after the efforts of the French women and the Chinese laborers, this group of soldiers comes together as in a quilting bee to piece together a localized patchwork. Indeed, even the simplest nets would be tailored around the specific natural or man-made landmarks they were sent to conceal. What is important to note here is that whoever installed the camouflage, whether an individual soldier or the infantry team as a group, had to learn to envision the context, to regard the immediate situation in terms of the absorption or reflection of light, not just as a defensive or offensive military position. Only in exceptional circumstances would infantry have actual access to recent, let alone real-time, photographs of their current environment.

Deceiving the interpreters of aerial photographs remained the paramount concern. As Caleb Arnold Slade, a *camoufleur* and pilot, explained in his 1917 "Notes on Concealment": "The aeroplane can reveal to the enemy range plotter as fast as it can photograph.... Diminishing the effectiveness of enemy aeroplane observations made from photos, camouflage is in its numerous ways falsifying the earth's appearance." As he continued:

> If camouflage is executed as ordinary landscape or scenic painting, its effectiveness would be lessened at a great distance of which it must be observed by the enemy.... Its effect must be duly studied at required estimated distance even as the paint requires certain distance for viewing. [It is] important to judge effectiveness at no other than the distance at which it is employed at the front.[63]

The *camoufleur* had to access a photographic point of view, to imagine what she or he was seeing not through his or her own eyes, but observed through someone else's. Thayer, for example, presumed that the lion saw the zebra standing in the reeds in the same way that he does and in the same way as the viewers of his cardboard

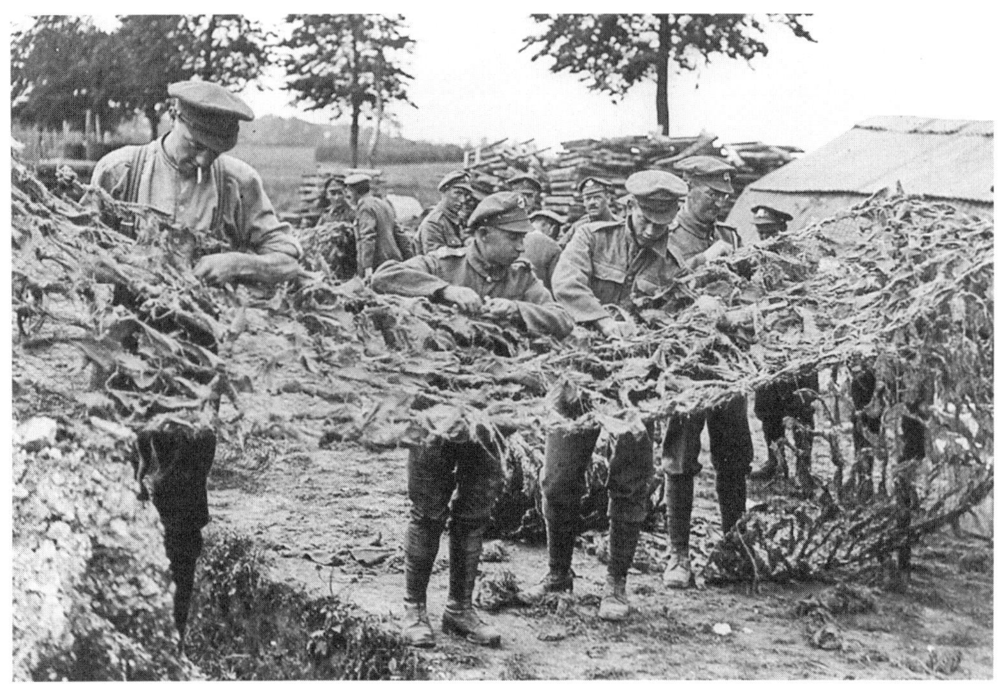

Figure 2.14—Mending the net. British soldiers near Besseux, France tie local foliage into netting sent from Special Works Park Headquarters, ca. 1917. Reprinted courtesy of the Imperial War Museum [Q6748].

models would. Here, the effective *camoufleur* had to work as if he were his own worst enemy.

As photographic emulsions improved, new filters made additional details of shape, shade, and hue visible. Camouflage technologies responded in turn, adaptively, so as to avoid detection by infrared, orthochromatic, and panchromatic films, as well as by a host of color filter overlays.

As the war dragged on and photographic technology achieved greater sophistication, the concealment of a group of objects—a dugout with soldiers or adjacent artillery dumps—increasingly required forms of netting that were both site specific and communally fabricated. A simple net such as the one depicted in figure 2.7 might have found its way onto the back of a howitzer, but the final customization—enacted by a collaborative team and incorporating a variety of textures, shades, and sheens—was always created in relation to a given environment and an assumed photographic apparatus. As aerial photography and camouflage rapidly codeveloped during World War I, optical technologies, natural landscapes, and human skills fused into the production of revolutionary photographic tableaux. By the armistice, a version of the L-type camera could make legible "images from 19,500 feet, using a four-inch lens to obtain crisp images in which even details such as barbed wire could be made out by intelligence."[64]

Textile Subterfuge

The World War I newsreel *Destruction of a German Blockhouse by a 9.2" Howitzer*, produced by the Army Kinematograph Company, depicts the preparation of a large gun for use just behind a trench line.[65] The key moments of drama, the first British attempt at filmic documentation of artillery in action, consist in the initial undressing of the gun (nicknamed "Mother") and, at the end of the film, "putting Mother to bed"—swathing the weapon in the sheaths of canvas camouflage. The ultimate goal, and the conclusion of the film, is concealment of the entire unit from the photographic eyes above and the cinematic eyes beyond.

Subterfuge such as this emerged as part of a dance between

innovation (aerial reconnaissance) and counterinnovation (camouflage). Participation in this dance of survival required constant recalibration in relation to changes in technologies of both viewing and killing, as well as actual or potential shifts in the enemy's position and point of view. In an aerial photograph, for example, a typical uncamouflaged trench line will appear as a white band with a black band running down its center. The whiteness is precisely what must be broken up through camouflage—creating shadows to mask the brightness. Solomon advised *camoufleurs* to "use netting to destroy sharp lines and angles—lay over installations to make them seem like smooth, sloping ground."[66] Meanwhile, the central dark shadow might be obscured by applying reflecting materials on the sides.

At every stage, camouflage development thus mediated between the exigencies of nature and the photograph; developing principles of site-specific optical concealment took experimentation and testing. Periodic testing and collaboration with photographic intelligence officers guided every material transition—from leaves, to canvas, to raffia, to scrim garnish, to netted gun covers and other netted camouflage forms.

While basic netting began with surplus fishing nets, sometimes hand knotted by fishermen or their wives, this was quickly revealed to be insufficient. Evenly garnished nets showed up as telltale "smudges" in surveillance photographs. Shadows and edges remained pronounced, despite a disruptive texture. Chesney early on noted the "increased incidence of such camouflaged positions appearing in aeroplane photographs."[67] The RFC and RNAS monitored the effectiveness of particular setups, and in 1917, the head of the Camouflage School specifically requested that airplanes from the RFC test new concealment measures.

Aerial testing revealed, for example, that while canvas strips faded less rapidly than did raffia garnish, they posed an altogether new set of problems. The canvas strips reflected large amounts of light, and thus the same color applied to raffia and canvas appeared in an aerial photograph lighter on the canvas than on the raffia. More texture was necessary; canvas was replaced with scrim, a lightweight

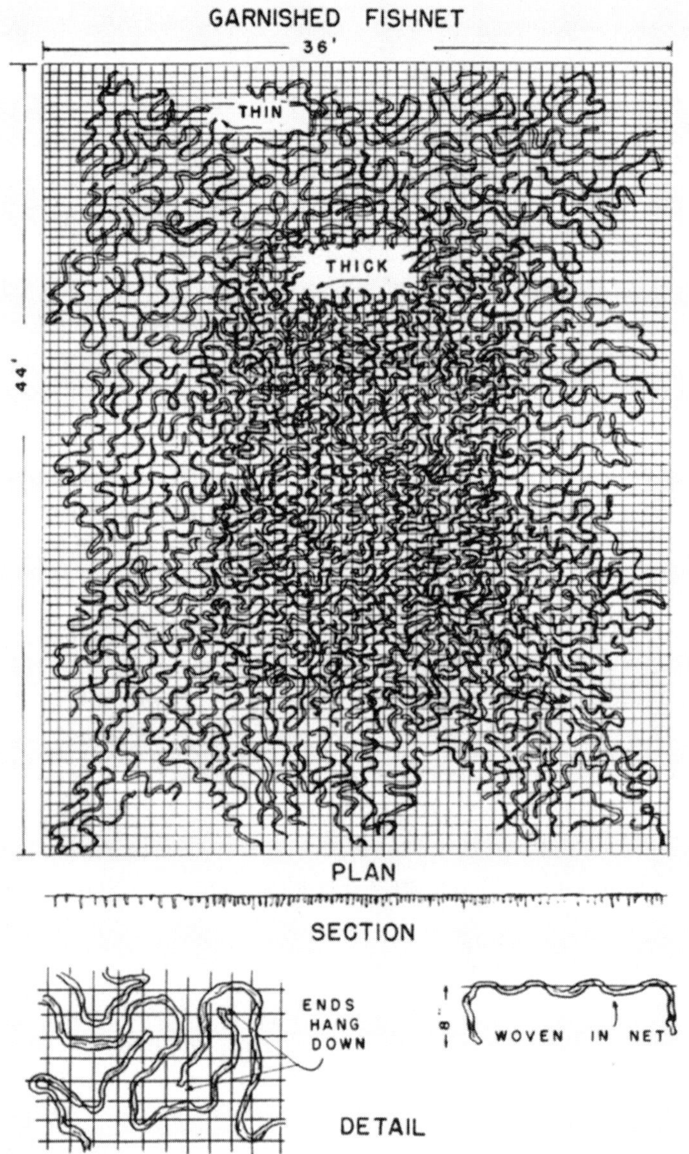

Figure 2.15—Garnished fishnet. Instructional diagram from the U.S. War Department's *Engineer Field Manual: Camouflage* (Washington: Government Printing Office, 1940), p. 12.

and semitransparent cotton veil. Photographic tests also demonstrated that nets could not be draped directly over angular objects and that sags would remain visible.

At the Special Works School in Kensington Gardens, aerial photography had revealed that nets garnished evenly also did little to diminish the shadows cast by objects underneath. Solutions included the issue of "prop sets" and adjustable, foldable poles to create smooth, naturalistic contours underneath the netting.[68] The problem of telltale shadows appearing in reconnaissance images of netting laid over howitzers was remedied by making the canvas strips or scrim thickest at the center of the net "and dying out towards the edges of the net," according to Chesney. "This improved matters very greatly."[69] The "flat-top" and related patterns were developed (figure 2.15). As Chesney remarked, "It was not till photograph after photograph had been exhibited, that they realized the vital necessity of irregularity for success in their camouflage work."[70]

This back-and-forth between photography in the air and craftwork on the ground culminated in the "sutured border" principle of camouflage. Fringed edges, such as those seen on the wings of butterflies, created a blurred effect, making interpretation especially difficult. Lieutenant Colonel Chesney referred to this strategy of rendering forms "hazy and very undefined" as the "sutured border treatment."[71]

Weaving the net, then, was a process of simultaneously getting "inside" both the aerial photograph and nature. It was found that "bright objects, such as packing-cases, ground much used and churned up by boots, cartridge-cases, etc., appeared in the photography quite distinctly through the net, so we provided uneven patches of scrim on the net itself towards the centre so as to render them more opaque in that area."[72] There needed to be nets within nets. Serial photography, the comparison of photographs taken in a predetermined sequence, was in Chesney's estimation the greatest threat to camouflage's effectiveness:

> With visual observation the object below is only glanced at, the point of view is changing, and changing very rapidly, so that it is difficult to

fix definitely in mind whether something we are desirous of finding is really genuine or merely camouflaged.... But with a photograph there is ample time to study it, to compare it at different periods of the day and of the week. Minute changes occur and it is these that are the clues.[73]

A coordinate system locates precisely any single entity within a frame. This fixed framework enables an interpreter to determine the path of an object in flux against what is assumed to be a fixed background—all that's required is a framework that assumes that change over time involves movement.

Netting responded to this coordinate system by acting directly on the emulsive boundaries of these photographic frames. Its secret presence confounded the detection of objects' movement via details of texture, shape, and shadow. It crept over streets and transportation systems, sometimes even before the roads had been laid. Whereas discrete or decoy objects became simply *other* objects of a visual field, the netting worked by actually occluding the field itself. As Chesney had described the "sutured border treatment," it fuzzed the photograph, as kudzu growing on a man-made structure eventually overtakes and conceals the structure around which it emerges. From the perspective of the outside observer, the structure has been reverted to nature.

In a logistical and tactical, as well as a metaphorical sense, netting seeded itself into the emulsive space both within and between the photographic frames. Netting was a veil that made it impossible to detect precise shape and texture—both in real time (by either aerial observer or cameraman) and in interpreted time (by the photographic analyst of periodized surveillance images).

Many hands fashioned these military mixed-media collages which, in situ, took on the form of site-specific and object-specific installations. This clothing was a weaving of natural and man-made materials that shielded military bodies of all shapes and sizes from photographic detection. Self-effacement was effected through projection into the natural environment, on the one hand, and through immersion into photographic emulsion, on the other.

Because "the material" was in a constant state of transformation, being a coproduction of state-of-the-art reconnaissance, site-specific technique, and the need to thwart the legibility and reliability of photographic evidence, the Allied armies lacked official guides and regulations on camouflage until well into the war. Not until 1917 did the Central Interallied Commission on Camouflage discuss developing common principles and production codes.[74]

Standardization followed. The Allied armies issued official camouflage guidelines to soldiers at the front in late 1917 and 1918. Authorities began to highlight the issue of so-called "camouflage discipline," individual behaviors that collectively could conceal the visibility of traffic flow and other large-scale military movements.[75] In the case of the installation depicted in figure 2.16, when combined with surreptitious behavior on the part of all troops, an aerial reconnaissance photograph (as opposed to this documentary photograph) would have ideally been duped by this dummy concealment measure.

In Saint-Gaudens's words, "to give the needed variation of light and shade, one had to blend the contours of the real (artillery, traffic route, salvage dump) into those of the ideal and peaceful landscape (forest, farmland, rolling hills)."[76] The former should be simultaneously immersed in and enveloped by the latter, but not in a static sense—not for the sake of "the photographic instant" only.

In Thayer's laboratory of the world, the instantaneous photograph ruled as arbiter of camouflage construction. A limitation of his project, as we have seen, was its insistence on a circumscribed field in which the viewer and the viewed—spectator and subject—were static, fixed in place. The photographer took the position of a predator and the model was prey, though both apparently were rooted to the ground. The instantaneous photograph rendered momentary concealment eternal.

With the advent of camouflage netting, the viewed had become active agents, operating to conceal themselves within regulated and serially photographed time, for the "material" was a skin in a kinetic and enduring sense. The viewer, meanwhile, was in motion, as "the machine" moved through the skies, on the one hand, and

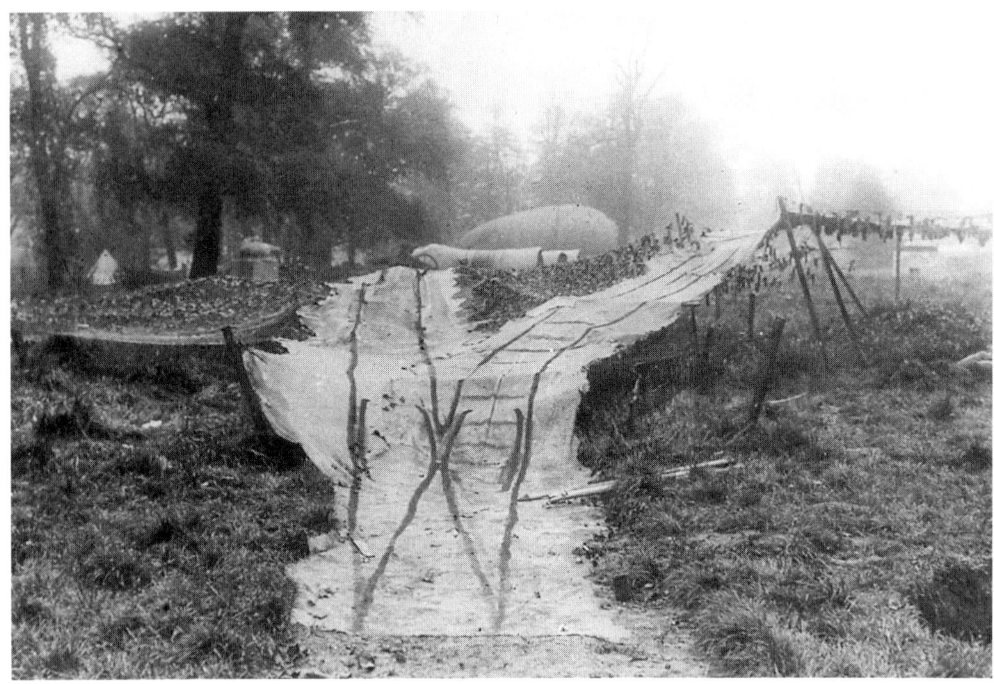

Figure 2.16—Dummy tracks. "Railway tracks" made of hessian canvas and black paint obscure a howitzer position; the "tracks" are extended with netting garnished with rags and leaves from nearby trees. Reprinted courtesy of the Imperial War Museum [Q17724].

as the interpreter analyzed a series of photographs, on the other. This movement on both sides necessitated that agents on both sides (ground and air) continuously projected themselves into the imagined position of the other. On the ground, one could hide only if one was able to imagine the view of oneself as seen from above. The *camoufleur* had to access a photographic point of view, even when physically based on the ground. The final customization—incorporating various textures, shades, and sheens—was created always in relation to a given environment and an assumed photographic apparatus. Camouflage became a mass technology, enacted on scales large and small; it was, to extrapolate from the words of film director and critic Jean Comolli, "the work of suturing, of filling in, of patching up the lacks which ceaselessly recalled the radical difference of the cinematic image was not done all at one go but piece by piece, by the patient accumulation of technical processes."[77] The *camoufleurs* had to fill in the lacks—insofar as possible—from what they saw around them. Only by this seeking could one keep up as nature marched on and the shutter clicked again.

Crafted by man, netting moved, as if with a botanical life of its own: this was textile subterfuge. As Colonel Slade noted: "The study of the locality for form and color that surrounds the objects to be camouflaged is necessary, and as everything is changing constantly—the seasons, the firing line becoming new environments—these things must be caught up with by change in the methods of camouflage."[78] The material clothed bodies as they shifted in shade, shape, and position, subtly and dramatically, too, through the changes of and in time, frame, and technology.

Large-scale technological advances during World War I precipitated a massive reconceptualization of self-concealment practices in relation to the photographic image. Individual and collective principles of camouflage discipline resulted in innovative forms of in situ concealment in which the hidden sought to see themselves as others might see them in the hope of disappearing. Abbott Thayer had shown how, frozen in time and space, the individual body could disappear into both the natural environment and the photograph

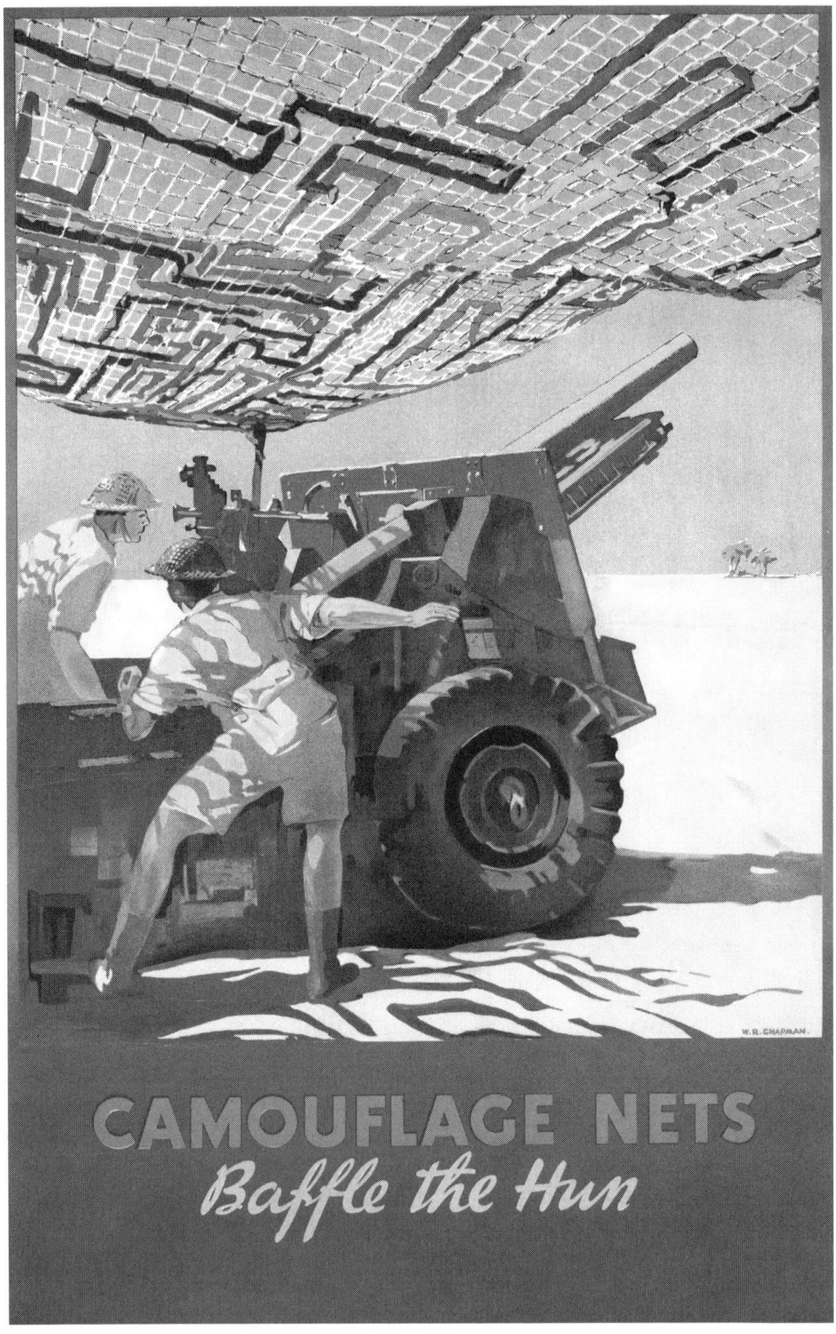

Figure 2.17—Camouflage nets: Baffle the Hun. This British poster was created by artist R. W. Chapman. The lithograph was sponsored and published by the Ministry of Supply and Her Majesty's Stationery Office. Reprinted courtesy of the Imperial War Museum [Art.PST 8034].

of that environment, and had advocated a discipline of seeing in which presence and absence collapsed into each other as in a photographic image. Whereas a single human body would have been unlikely to appear in an aerial surveillance photograph, tracks of movement through a landscape—be they tracks of a single human or of a collective—might well show up on film. Under the pressures exerted by the serial photography of aerial surveillance, the need to "cover one's tracks" in the most literal sense produced new ways to disappear and, with them, new ways to locate the self so as to hide in plain sight.

New surveillance technologies continued to proliferate after the Great War ended, and with them came not just new camouflage technologies to match, but a heightened sense of "camouflage consciousness" (figure 2.17). It increasingly became necessary to learn new forms of bodily movement, as well as ad hoc bodily adornment, so as to blend in while simultaneously projecting oneself into the position and perspective of someone who might be lurking, and filming, anywhere.[79]

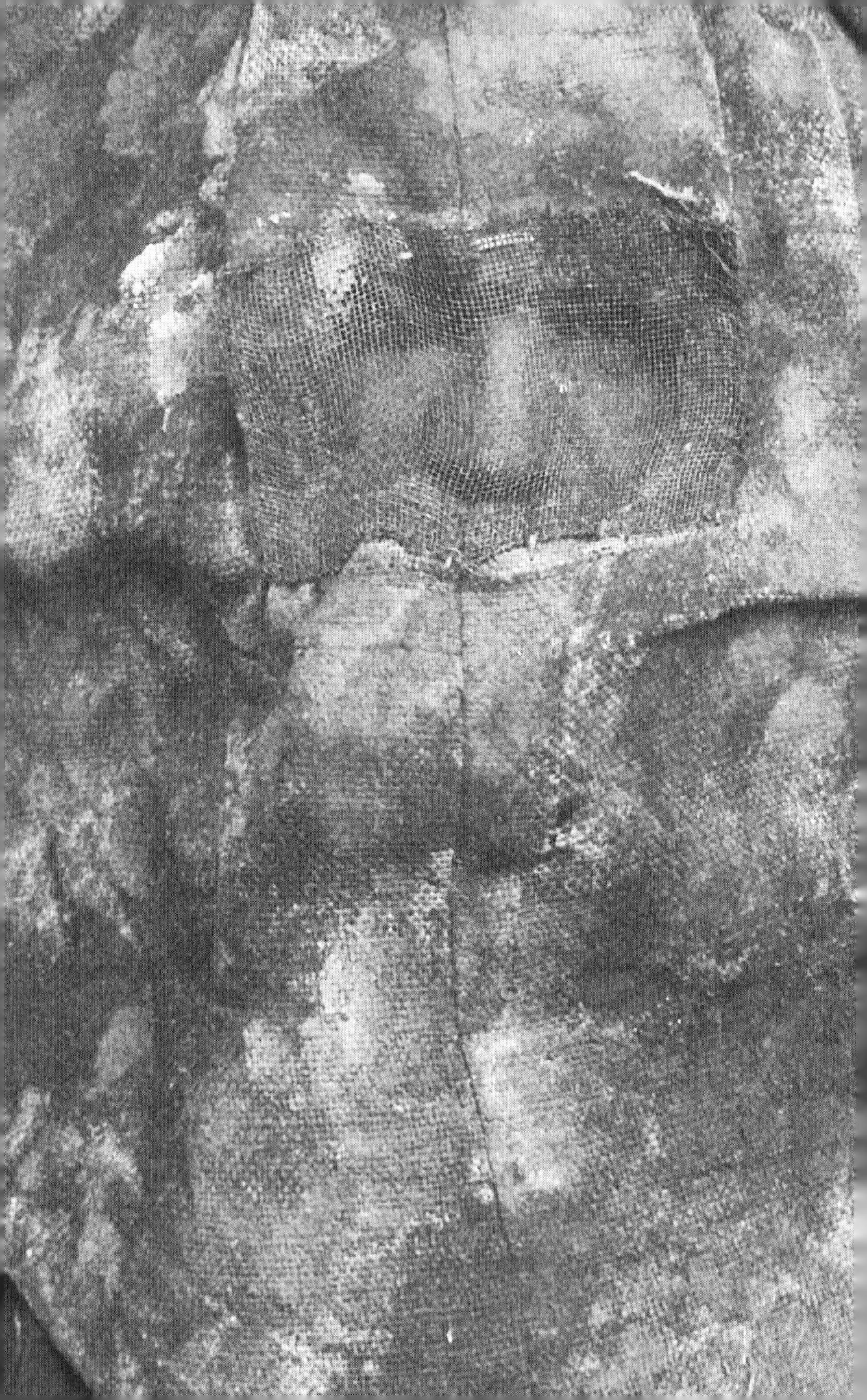

CHAPTER THREE

How Not to Be Seen

Dynamic Camouflage

and the Motion Films of Len Lye

In the Monty Python skit "How Not to Be Seen," or "H.M. Government. Public Service Film #42, Paragraph 6," a disembodied narrator voiced by John Cleese instructs viewers in "how not to be seen." The film version begins: "In this picture there are forty-seven people. None of them can be seen. In this film, we hope to show you how not to be seen."[1] Viewers are presented with successive views of rural landscapes and alerted to the existence of an invisible subject camouflaged somewhere within each. The narrator, however, eventually identifies the whereabouts of the human subject within each landscape. Each exposure is followed by a sudden explosion of the exposed. Audience laughter inevitably ensues. As the spoof educational film draws to a close, the disembodied voice-over source materializes, and John Cleese sits at an oak desk positioned in an open field. The camera zooms in as he cackles convulsively. Then he, too, is blasted into oblivion.

The supposed instructor (Cleese) shows great skill at identifying others, but fails grandly at self-concealment. The irony is fittingly explosive and profound. In the skit, the only subject, and indeed the only object implicated by the film that actually knows and performs "how not to be seen" is the viewer—you. By maintaining a position of pure spectatorship on the "other" side of the screen, the viewers, initially posited as the students, know best. We may or may not choose or be moved to laugh, cry, or do something else entirely, and in this sense, as the unseen, we are ambiguously and freely positioned in the scene.

And yet, outside *Monty Python's Flying Circus* and outside the comfortable space of the darkened civilian theater, the human subject of cinematic instruction is—like a military sniper in a war zone—often already an object of photographic surveillance. The Monty Python skit inscribes a potent reality. In science, nature, war, and documentary filmmaking alike, learning the art of "how not to be seen" is not funny at all. Rather, it is part and parcel of a certain kind of learning how to see, and how not to be seen, in nature and in relation to the camera (figure 3.1).

The Monty Python skit is a parody of World War II instructional films produced to train British civilians how better to survive a Nazi invasion. Fear that the British Isles would be invaded had been widespread ever since Germany's declaration of war in 1939; many civilians thought it imminent. The military worried especially about less developed areas—especially areas in the more remote rural regions of England, Scotland, and Ireland. The British War Office established a unit to help collect and train civilians in military preparedness. Over the course of the war, more than 1.5 million men who were otherwise ineligible for normal military service volunteered for this group, initially called Local Defence Volunteers (LDV). Nicknamed "Look, Duck, Vanish," the division was officially renamed the Home Guard in November 1940.[2]

LDV enlistees required training first and foremost in principles of offensive and defensive subterfuge. If and when the Nazis surreptitiously airdropped snipers and spies onto British soil, the locals had to know how to "see" them (snipers) and how to "see through them" (spies under cover).[3] Training the LDV presented an enormous challenge, because resources were extremely limited; there were nowhere nearly enough weapons or experienced instructors to go around. Officials in London decided to try using films, which could be easily distributed around the isles via mobile cinema units and screened in local theaters, churches, schools, and libraries.[4]

The Ministry of Information (MOI), founded in September 1939, oversaw through its in-house Film Division the production of instructional films on topics ranging from visual training, to

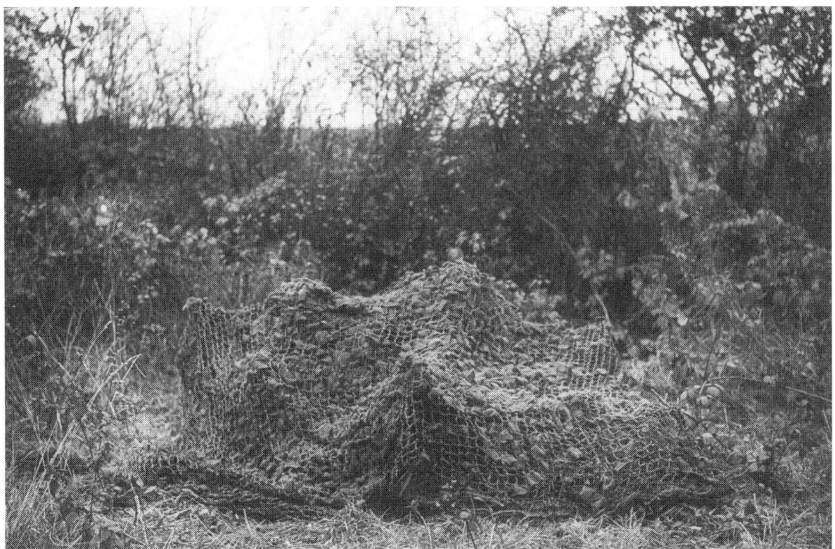

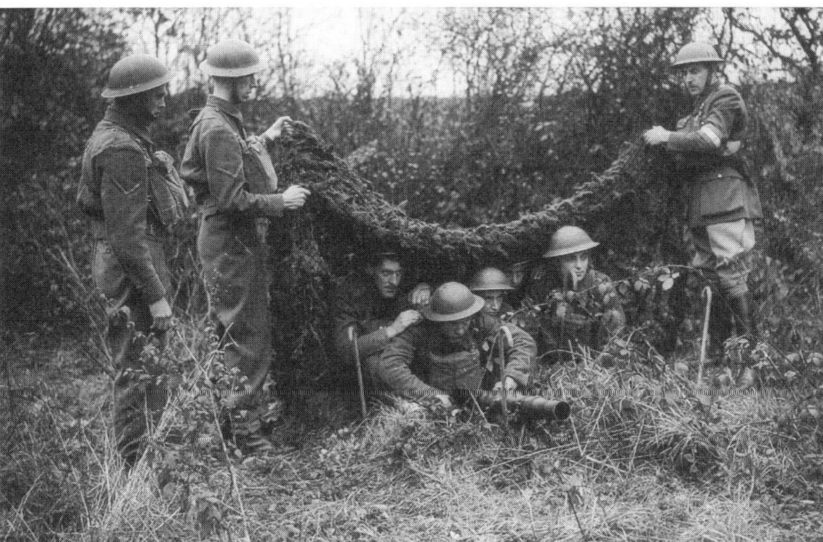

Figure 3.1—**How not to be seen.** British soldiers, seen here in both undercover and exposed positions, demonstrate effective self-camouflage in these ca. 1941 instructional photographs. Reprinted courtesy of the Imperial War Museum [H5465 and H5467].

marksmanship, to how to scan the sky for airplanes, all seen as relevant topics, given the felt threat of Nazi invasion. The MOI also sponsored the production of realist wartime features and shorts whose intended purpose was to boost morale and preparedness among the civilian population.[5] Some of the films never made it out of the editing room; some never left London. Others were screened throughout the country, during and after the war. Some of these, according to viewer reports, provoked yawns, while other films had audiences on the edge of their seats. One of the latter ended up crossing both genres and oceans (figure 3.2).

In August 1943, *New York Times* film critic Thomas Pryor announced the arrival of one such Ministry of Information film in U.S. theaters: "an especially exciting and brilliantly executed piece of pure cinema which goes by the title *Kill or Be Killed*. This is the true story of the stalking and counter stalking of a British (sergeant) and a Nazi sniper. For eighteen tingling minutes, the two snipers, resorting to various tricks of camouflage, hunt each other through open fields and woodland."[6] Pryor described the film's ultimate effect in sonic terms. "The crack of a twig inadvertently stepped upon rings out in the deadly suspenseful silence." *Kill or Be Killed* was, according to the press release, "a story of a man-to-man battle of wits between an English Sergeant and a German sniper."[7] An earlier review in *Documentary News Letter* praised the finished film as "without a doubt one of the most exciting ever made."[8] (Another reviewer was more blunt in his appraisal: " manhunt."[9])

Kill or Be Killed was directed by avant-garde animation and kinetic artist Len Lye and his team, working out of the Realist Film Unit (RFU), a nonfiction film production house founded by documentary filmmaker and critic Basil Wright. It was first released to the Central Film Library for nontheatrical distribution in January 1943. Within a few months, it was being shown to employees of the Home Guard and British Civil Defense throughout the country, part of the traveling series organized by the Film Division's mobile cinema program.[10] But theatrical audiences were soon identified, as well, both in Britain and in the United States.[11] Its narrative and

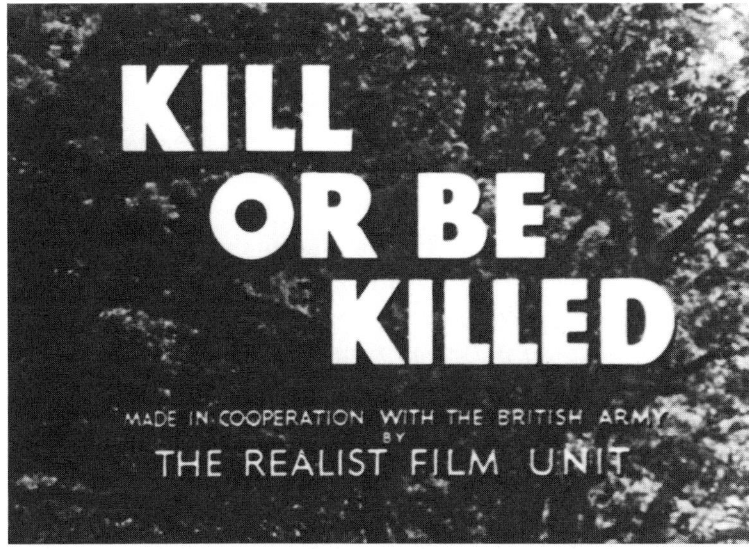

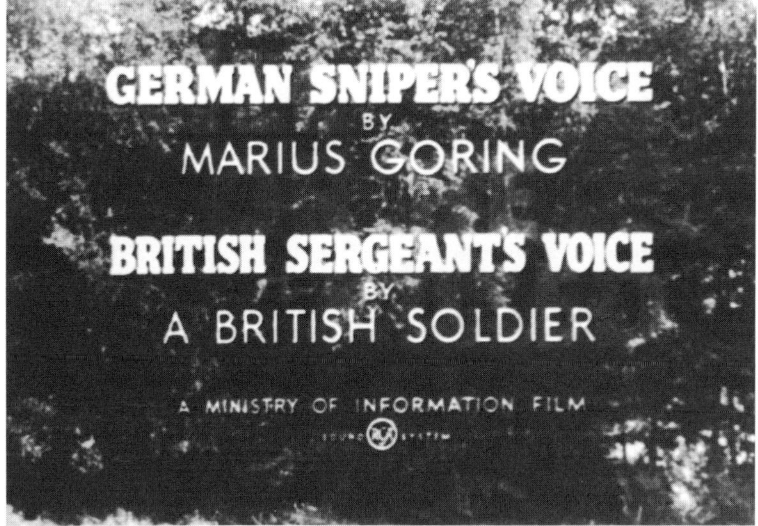

Figure 3.2—Kill or be killed. Stills from the title sequence of Lye's film. Reprinted courtesy of the Len Lye Foundation, Govett-Brewster Art Gallery.

aesthetic structures were unusual among training films of the era. It was nonexpository, and, as the *New York Times* described it, there was "no dialogue except for an occasional voice representing the silent thoughts and fears which run through the soldiers' minds."[12] Its unusual structure and aesthetic worked on its audience subtly; it "has the magic of cinema in it, every legitimate device has been used with an integrity which is rare in film making. *Documentary News Letter* concluded its review in more direct and utilitarian terms: "Propaganda value: Excellent."[13]

Close analysis of *Kill or Be Killed* in the context of the historical moment of its construction and viewing, reveals considerably more. *Kill or Be Killed* was an emergent multimedia practice of filmmaking (and film viewing) that remains with us today. As we have seen, camouflage had a symbiotic and combative relationship with filmic image-making practices. As the *New York Times Magazine* reminded the public in 1941: "The camera makes its record in black and white—in tell tale shadows. And the great task upon which camouflage is employed is contriving means of making the camera's record in enemy hands a false report of the facts."[14]

This emergent form of multimedia practice, which I call "dynamic camouflage" and which emerges in Lye's work, designates both a survival strategy focused on protective concealment of the mobile body within a changing environment, itself under filmic surveillance, and a formulation of subjectivity, a form of consciousness adapted to that environment. It is founded on a specific logic of self-effacement that collapses the distinction between the filmed world and the natural (or protofilmic) world. As embedded in *Kill or Be Killed*, dynamic camouflage can be understood in narrative, aesthetic, and logical terms as a model of embodied pedagogy, that is, as an aptitude both incorporated in and expressed by the body, rather than as mere "tricks of camouflage," to quote the *New York Times* film critic. Movement of the body increasingly had to be orchestrated with avoidance of indexical photographic registration in mind, a response to continued optical surveillance in which the position and movements of the camera are unknown to the observed.

Whereas Lye's film stood out from the audio-visual media environment of its own day, it blends in surprisingly well with our own. The conditions under which dynamic camouflage could emerge included both the threat of optical, real-time surveillance from an unknown position and the possibility of an embodied and mentally engaged subject operating in a natural environment with which she is able to interact and intervene, bringing together found and fabricated materials in the service of immersion and disguise. Camouflage fieldcraft and photographic and cinematic surveillance technologies became interwoven concerns in the years leading up to and including World War II. On the one hand, photographic reconnaissance increasingly became precisely the visual medium from which one in danger sought to efface oneself; people needed to learn how to use fieldcraft to hide from such surveillance.[15] On the other hand, film was becoming a more and more prevalent means of military training. The question was how film itself might be used to provide effective instruction in how to do so. In *Kill or Be Killed*, Lye provided an answer to this state of affairs. His work underwrote a new trend in visual aesthetics and media practice whose remnants have since been absorbed into media producers' practice more broadly.[16] Certainly many other aesthetic sensibilities can be seen as precursors to the first-person-shooter ethos that marks both recreational gaming and military training devices in the twenty-first century. But the structure of Lye's *Kill or Be Killed*, taken in its context, introduced a vision and made manifest the conditions for the emergence of the media environment we move through, literally and subjectively, today.

Experiments in Mimicry

Lye brought an unusual background to the production of a pedagogical film about civilian concealment in wartime; he was no ordinary military man. "Len Lye could be described in the history of British cinema by one word—experiment."[17] So reported Alberto Cavalcanti, a filmmaker and Lye's sometime producer. For New Zealand–born Lye, kinetic motion inspired an extended period of aesthetic and philosophical investigation prior to beginning to work in live-action public-service film.[18] His nondocumentary background—his attention in particular to the materiality of the filmic medium—affected the approach he took while working with Realist Film Unit.

Soon after arriving in London as an indigent artist in 1929, Lye made *Tusalava*, a five-minute abstract animation piece. Members of the exclusive London Film Society took an immediate shine to its organic, almost biomorphic, quality, and Polynesian motif. Lye scratched and picked and etched onto his films, a process that came to be called direct animation. (More generally, direct animation refers to techniques of making images directly on film stock, without the use of frame-by-frame photography.) With the successful reception of *Tusalava* at the Film Society, Lye had entered the London filmmaking scene, an arena wherein experimentation was welcome. The Film Society indeed actively brought together those engaged with what we might in retrospect categorize as avant-garde, commercial, and educational filmmaking; in this period, these circles were tightly connected.[19]

Despite the exclusivity of their screenings, the Film Society's directors did not shy away from broader public engagement, and in 1933, John Grierson, who for a time ran the Film Society, inaugurated a sponsorship program with the General Post Office that would enable the production of animations, puppet pieces, and other short works. Those involved in the productions of the GPO Film Unit were a motley bunch of poets, writers, artists, and camera operators, and Lye joined in under Grierson's tutelage. At the GPO Film Office, Lye pioneered the application of tools—paint, stencils, cutouts, solarization, and batik-derived motifs—to emulsion. Lye's early filmmaking experiments in direct animation thematized color change over time through distortion and erasure of recognizable figure-ground relations. Unlike much of the painted film that was wildly popular earlier in the twentieth century, Lye did not "color in" images depicted in grayscale on the surface of the film. Rather, he colored *on*, and injected pigment *into*, the celluloid emulsion itself. As Lye described direct animation practice, "It is the means by which you directly etch, that is, scratch with a needle, right into the celluloid, or paint right onto celluloid so that the color sticks to it."[20] He treated the film as the recipient of pigment, applied like makeup, and in this sense, Lye resembled a tattoo artist whose skin of choice was film. The designation of film as skin can be taken literally here. Both skin and film function as interfaces and border zones between organisms and their environment.[21]

Lye explained the motivations behind this practice as arising from a desire for greater control. "I've been reduced to painting on film, or scratching film, not that I want to, but because I want to deal with the control of color and three-dimensional motion," he told two film curators of the Museum of Modern Art years later.[22]

Some of Lye's experiments bore strong conceptual and material links to the static and serial modes of camouflage explored in the previous two chapters. In *Colour Box*, made in 1935, Lye experimented with painting, scratching, and pasting stenciled shapes onto discarded pieces of film, creating a five-minute-long sequence of dynamic color. Lye then optically printed the painted film using the Dufaycolor method, which transformed a collage of painted emulsion and splices into a continuous band of film, creating a palette of soft, greenish blues and muted pinks in the process.[23] Setting the stencil into motion prefigured Lye's eradication of figure-ground distinctions in subsequent films as well. When projected, "pure motion," in Lye's words, should replace continuity of form.

Circular forms in the shape of holes mark both the composition and rhythm of Lye's films. In *Colour Box*'s opening sequence, dancing forms fuse with metamorphosing colors, and movement simultaneously guides and masks the transition between frames. Shapes cut into the moving background seem to fuse and diverge as their position progresses across the frames. Serial photographic strips emerge, via projection, in the form of visually continuous motion. Lye furthered his interrogation of motion by incorporating into his composition the sprocket holes that line motion picture film's top and bottom edge. These sprocket holes, cut into pieces, actually became the stencils. In live-action film, such sprocket holes—the functional core of the cinematic apparatus—are invisible to the viewer of the projected cut. But in Lye's pieces, as in Abbott Thayer's stencil constructions, holes become forms. Absences are transformed into pigmented presences (figure 3.3).

Lye also highlighted the application of stenciled "sprocket holes" and silhouetting in his five-minute animated Technicolor film *Trade Tattoo* (General Post Office, 1937), which made more explicit reference to the permeability of sixteen-millimeter film material in relation to transformations in both color and sound.[24] The very word "tattoo" simultaneously carries military, pedagogical, painterly, and epidermal overtones. In sonic terms, tattoo (derived from "tap toe") refers both to a signal drumbeat that calls soldiers to order and to a similar drumbeat soldiers create for their own entertainment. In both cases, a drumbeat impels bodily movement; rhythm motivates soldiers, and discipline is triggered by access to a deeply felt rhythm. In the epidermal sense, "tattoo" (derived from the Polynesian *ta'tau*) refers to the permanent stain or design of skin caused through puncture practiced by seamen since the eighteenth century.[25]

Lye's *Trade Tattoo* was produced by acts of collage—manipulation of black-and-white documentary photographic footage layered into a form of disguise. In addition to painting and drawing over their frame lines, Lye used outtakes from classic GPO documentaries (mostly from Basil Wright and Harry Watt's 1936 *Night Mail*) to construct filmic collages.[26] He printed exposed and unexposed, perforated sixteen-millimeter film on top of footage of British seamen engaged in shipping goods. Using both found and fabricated materials, then, was as central to his practice as it was to the *camoufleurs* discussed in Chapter 2; embedded in serial reconnaissance photographs were the traces of acts of subversion, attempted trickery of the photographic interpreter by the groundlings in action. The purpose of these enactments of camouflage was to disguise vital movements and positions and reveal false movements and positions in their stead. In *Trade Tattoo*, Lye similarly applied patterns and stencils to flatten out objects against their moving

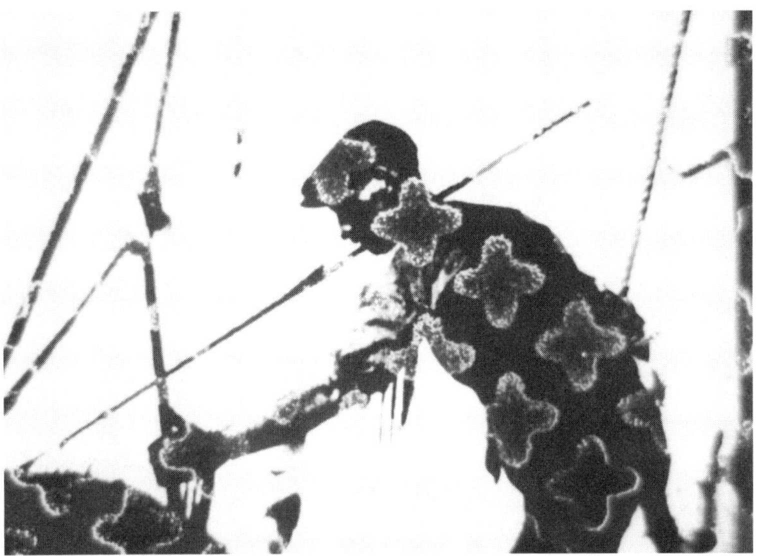

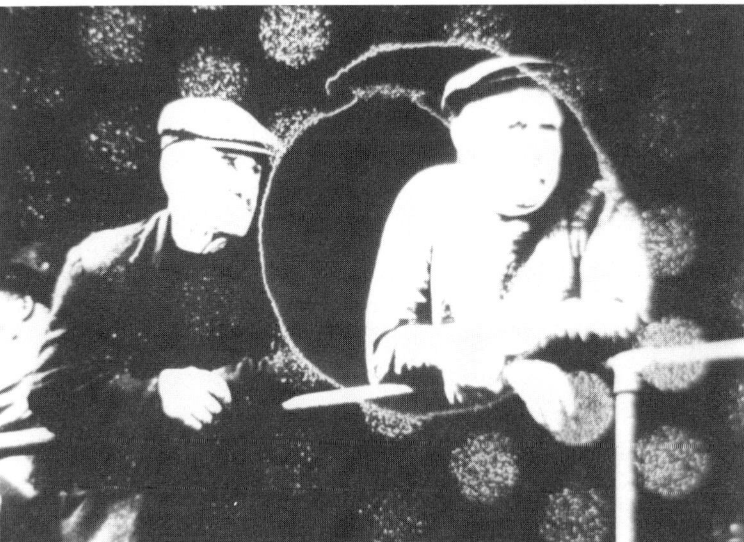

Figure 3.3—Acts of collage. These stills from Len Lye's short film *Trade Tattoo* (1937) exemplify key aspects of his approach to experimental animation. Lye disruptively transforms the photographic document through complex manipulation of color, application of layers of stencils and scratches to the filmic substrate, and optical printing. By these means, he simultaneously effaces the supposed "truth" of the documentary footage, and calls attention to precisely these acts of manipulation and subversion. In the top image, a sailor has been dressed in a disruptive pattern material, rendered in hues of chestnut brown, hunter green, and olive drab. Reprinted courtesy of the Len Lye Foundation, Govett-Brewster Art Gallery.

backgrounds. The effect is of a shape-shifting disruptive pattern material, a concealment of the human referents to the live-action footage beneath.

In his particular rendering of documentary footage, the finished film is itself a patchwork quilt of textile and body-art practices. Textile traces are evident in *Colour Box* as well as *Trade Tattoo*. These direct animations become pieces of dynamic art in motion—batik fabrics billowing in the wind, color and movement mimicking each other as they are squeezed together.[27] As Thayer would have construed it, the living subject (human) and nonliving background seem to be cut from the same cloth. Now, however, in the realm of the dynamic, the subject and background were both in motion.

In the case of *Trade Tattoo*, this transformation makes the original human subjects of the found footage appear to disappear. But this "dressing up" of the film is also an operation that conceals its material and photographic origins. It connects otherwise discontinuous live-action footage and papers over foreground-background contrasts, just as it papers over the original subjects that dominated the live-action footage substrate. In Lye's animation, documentary film itself was erased by the genesis—through movement—of color and organic form.

In these works, then, film became akin to a second skin through which the origins of the work of art itself could be obscured. Stills from Lye's *Trade Tattoo* recall the textile patterns of experimental camouflage patterns from World War I, versions of which were beginning to be standardized by the British and American armies by the first years of the World War II. Artist-innovators, including Abbott Thayer, Solomon J. Solomon, and Homer Saint-Gaudens, proposed sniper uniforms woven of contrasting shapes and hues. Cornwell-Clyne advised in his "Notes on Concealment": "It is not a difficult matter to make a head harmonize with the foreground. For example, in a turnip field you must look as much like a turnip as possible, in a cabbage patch like a cabbage, in pasture like sod."[28] It was hoped that such measures would break the distinction between the human body and its environment. Enemy sights should detect only a generalized motion in which colors and shapes moved without delineating the precise form of the wearer.

Filmcraft as Fieldcraft

Lye's inventive style as animation and textile artist would affect his work in live action. In May 1942, the Ministry of Information commissioned the Realist Film Unit, based in London's Soho, to prepare a treatment and script for a film on strategic concealment. Realist was one of about ten main suppliers of film to the Ministry of Information. The target audiences for this film would be both soldiers and civilians.[29] Realist put Len Lye in charge of the project, which the Ministry of Information planned to incorporate into its "mobile cinema" program. The MOI hoped that this, among other wartime-preparedness training films it sponsored in 1942, would shortly be screening on makeshift screens erected in factories, churches, schools, and hospitals.[30]

At the very beginning of World War II, the British Army had established a new Camouflage Development Research and Training Centre, modeled on its World War I predecessor, the Special Works Park, as well as a related intelligence unit, the Special Operations Executive, or SOE.[31] In contrast to the emphasis on surveillance, and particularly aerial surveillance during World War I, individual camouflage practices during World War II were increasingly entangled in discussions of "fieldcraft," defined in 1942 as "the training of soldiers in stalking, crawling and patrolling, as well as strategy and tactics for hiding and seeking in a dynamic terrestrial environment." It described principles required to operate stealthily by day or night, rain or shine: Fieldcraft is the art of the hunter, coupled with the wiles of the poacher. "This includes concealment, silent movement, knowledge of his prey and skill with his weapon; also the use of both natural cover and camouflage in conjunction with movement."[32]

Successful fieldcraft also required training the self to adapt constantly, materially, visually, and even psychologically, to its surroundings. Destroying one's enemy depended on it. As another 1942 handbook on fieldcraft put it: "One's brain must be constantly active and on the alert—thinking ahead, so to speak: taking advantage of ground and cover, light and shade, wind, noise, rain, mist, snow—ever on the alert."[33]

Maneuvering in such an environment demanded perceptual acculturation and training in walking and turning. Different types of crawling, for example, were suited to specific environments and to the specific natural resources at hand—be it underbrush, ferns, mud, gravel, or manure.[34] These materials could also be mixed with canvas strips, other garnishes, shoe polish, face paint, or rubber bands to make oneself up. Creative use of features of the natural landscape (leaves, branches, mud) and textile materials (rags, raffia, painted canvas)—"ecological collage," one might call it—was essential. "Use the country; become part of the earth upon which you walk or lie or hide; make yourself invisible with leaves, or earth stains, or with lightly teased strips of bark."[35]

The hunter-hunted dynamic was an old one. Some of these practices had been deployed in a British military context since well before the Boer War. In the realm of hunting, such principles had been developed among the ghillies, or game wardens, of Scotland, for centuries.

Blending into the woodwork when the woodwork itself is constantly changing was no easily practiced or filmable task. In the domain of static camouflage inquiry and representation and in the lives characterized thereby, one needed to be an invisible statue—the ghostly element of a perpetually static picture—for an instant and only for the purposes of documentation (real or imagined) in a single still photograph. To cite Thayer and his son: "It is for these moments that their coloration is best adapted, and when looked at from the viewpoint of the enemy or prey as the case may be, proves to be obliterative."[36] In the context of fieldcraft and in the representational realm in which Lye practiced his filmmaking, the goal was seamless movement. One had to approximate a stencil, along the lines of Thayer's stencils, but a stencil whose borders were fluid.

The cultivation of good fieldcraft was a lofty goal for audio-visual education—be it slides or films. Learning how to disappear as a kind of second nature seemed to exceed the possibilities of standard military drill routines or standard 1940s instructional films.[37] Traditional educational training films were expository, involving a direct audience address by a voice-over narrator.[38] The Monty Python skit "How Not to Be Seen" makes a mockery of precisely this kind of approach to the subjects. In these training films, the narrator provided instruction on how to perform a certain behavior or set of behaviors (loading a gun, for example). A generalized "self"—whom I will call the "model person"—performed the task on-screen, fully visible to the camera and the viewer alike. The viewer, so invoked, would occupy the position of the viewfinder, ordered to follow the behaviors of the model person. Delayed imitation of another person's specific behaviors and gestures—this was the model. In the British context, films such as the 1943 *Camouflage and Fieldcraft* series,[39] produced by the Army Kinematograph Society, adopted this approach, as will be discussed later on.[40] With the aid of the narrator, the model person served as a direct illustration of the lesson at hand.

In the case of camouflage, the limitations of this approach are evident. How, after all, do you show somebody how not to be seen? That it seemed to strain the possibilities of standard military classroom pedagogy is a point made strongly by the Monty Python parody. John Cleese's explosive end suggests that the ability to spot—or mimic—a person attempting invisibility does not ipso facto lead to a talent for self-concealment. Which isn't to say that many hadn't tried; a series of slide lectures circulated widely as part of a broader audio-visual program of education for soldiers, who would see still images projected on-screen while text was read aloud.[41] And yet learning how to disappear as a kind of "second nature" required more than glass lanternslide lectures, textbook instruction, or standard voice-over training films. Modern warfare required a "camouflage consciousness"—less a matter of mimicry than a function of immersing oneself into a filmic understanding of one's self moving against a dynamic backdrop.

Lye brought his interest in experiments in "pure motion" to the task of camouflage filmmaking. In the treatment and script approved in July 1942, Lye outlined a psychologically complex, character-driven, cinematic "lesson in camouflage, a lesson in stalking," that would come to be known as *Kill or Be Killed*. Lye cast the film's two main parts with a civilian ghillie for the part of a British sergeant named Smith and with a British stage actor known for his previous work impersonating Hitler in anti-Nazi radio broadcasts for the part of the Nazi sniper, Schmidt. Lye collaborated with cinematographer Adrian Jeakins and second cameraman John Harding, and, together, they shot the film on black-and-white thirty-five-millimeter stock in Chobham Wood, Surrey, in July and August of that year. Edited in September, the result was a sixteen-hundred-foot, eighteen-minute film. Sixteen-millimeter prints, as well as limited thirty-five-millimeter versions, were released soon thereafter.

Lye submitted the film to the Ministry of Information in November 1942. Within the UK, it was projected in schools, factories, and libraries in front of assembled civilians from many walks of life.[42] Soldiers did watch the film, not so much as official training, but more as a kind of complementary entertainment. Some reviewers hailed *Kill or Be Killed* as a "masterpiece of filmmaking." Producer and filmmaker Basil Wright, cofounder of the Realist Film Unit, wrote to Lye in November 1942: "Heartiest congratulations on kill or be killed which is one of the most exciting films and one of the best jobs of film making I have seen for a long, long time."[43] Others worried that the film struck a nerve with rural and civilian audiences, but for the wrong reasons. Roger Manvell, a historian and Ministry of Information producer who would cofound and direct the British Film Academy in 1947, highlighted *Kill or Be Killed* in an essay on the relationship between realism, shock, and humor in documentary films of war. Manvell was distressed by the "titters of uninformed audiences" when violence erupted on-screen. The laughter was "due to mild shock," he felt, while "severe shock, such as the actual experience itself, would lead either to hysteria or greater self-control according to temperament." In a similar fashion, the language of the film prompted Manvell to dwell on the dissonance between the horrors of the situation depicted on-screen and the knowledge that one existed "safely" on the other side of the screen:

> The evident intention of this film is to be realistic within the framework of an uncensorable treatment. The language is strong, but less strong than the situation would actually warrant; the subject is as grim as anything in war can be: two soldiers, one British, one German, manhunt each other. Barring details produced by captious criticism, there is nothing in the film which could not happen. And yet, when the British sergeant, having barely escaped with his life from the marksmanship of the German sniper, says "I'll get that Jerry bastard before he gets me" (a perfectly reasonable remark) the audience shouts with laughter, or gives a mild titter if it is drawn from the more country set-up. Thirty seconds is then required to restore the atmosphere intended by the situation as a whole.[44]

What was so exciting, singular, and powerful about *Kill or Be Killed*? It captured fieldcraft and transmitted it to film. Its potency and military usefulness was genuine, even if not explicitly or immediately intended for infantrymen. Through its narrative, pedagogical framing, and aesthetic approach, the film conveyed a logic and aesthetic of camouflage. Nature, film, and war fused in dynamic practices of self-effacement.

Structures of Self-Effacement

The structure of *Kill or Be Killed* can be broken into three parts, or acts, moving back in time and then forward again. Act 1 depicts the death by shooting of a German sniper, Schmidt, by a camouflaged British sergeant, Smith. Act 2 is an extended flashback; the previous five hours are compressed into ten minutes as the sergeant shifts from a position of vulnerability to one of ascendancy. His bodily exterior and pattern of movements are transformed according to the principles of dynamic camouflage. He immerses himself in his natural environment, at the same time projecting himself into the position of a potentially all-encompassing photographic surveillance device. Fieldcraft is transformed so as to embody a response to photographic surveillance and to filmic observational practices more generally.

Act 3 returns to the opening scene; the fatal shot rings out, and the Nazi is killed. As the flashback melts into the present, a more nuanced future follows in real time. The final three minutes of the film depict the construction of a dummy out of the dead body and the appropriation of the Nazi sniper suit for the Brit's own use. The film ends as a fleet of aerial reconnaissance planes, engaged in their own surveillance mission, flies across the screen overhead. Ultimately, the flashback structures the film in terms of a projection of the self into the position of the model person, on the one hand, and into a position where one's own activity over time is visible to what New Zealand sniper Ion Idriess, author of the manuals *Sniping* and *The Scout*, would call "the mind's eye." As Idriess advised in *The Scout*: "See the enemy with your mind's eye. The scout must see, but not be seen. He must hear, but not be heard."[45]

Over the course of *Kill or Be Killed*, the viewer might, according to her disposition, experience the thrills of both hiding and seeking simultaneously. Through an oscillation between distinct points of view, she alternately occupies the positions of sergeant and sniper (figure 3.4). Identification with the model person (the English sergeant) and then with the model enemy (the Nazi sniper) is followed by immersion of the sergeant into his technonatural, mediated environment and of the viewer into the position of camouflaged filmmaker. Inscribed into the film is a performative logic for viewer, filmmaker, and soldier. Seeing comes to be the result of manipulating one's environment so as to remain unseen.

The twenty-second-long opening shot of *Kill or Be Killed* begins as a pan downward. The pan settles, and the camera focuses on a field, a break between two wooded areas. A previously concealed man in the lower left corner begins to wriggle. He writhes his body, inching slowly forward as he rights himself into position. It is this act of wriggling that brings him to our attention. So objectified, the model person becomes a specimen—an insect, as it were—to be identified and then analyzed, as if by a naturalist. The viewer is positioned as the

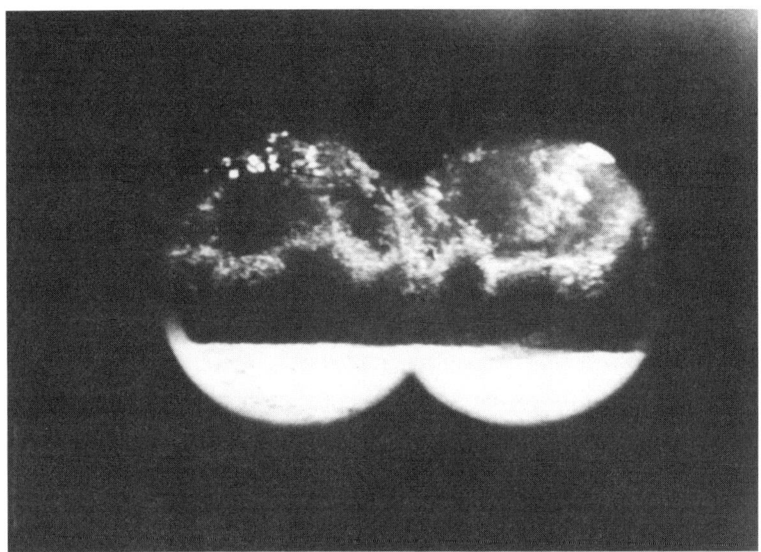

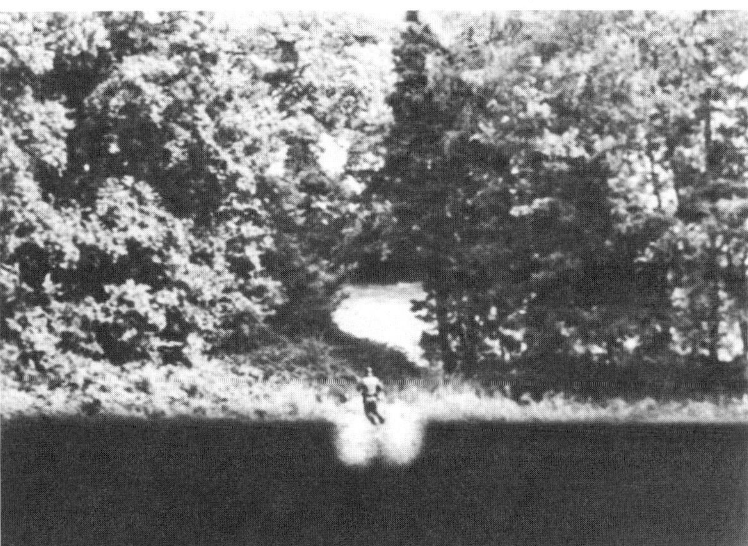

Figure 3.4—Oscillating points of view. The film viewer shifts between direct examination of the filmic field through German binocular lenses and British rifle sights, among other first-person photographic surveillance devices over the course of *Kill or Be Killed*. Reprinted courtesy of the Len Lye Foundation, Govett-Brewster Art Gallery.

detached observer of a visual target (the sergeant) who himself seems to have identified his own target (the sniper). This is the point when, were this a Monty Python skit, the crew would detonate the now-visible target. Neither the filmmaker nor the film viewer is among the sergeant's worries, however; he does not appear to anticipate their presence, let alone to hide from them. A fateful bullet is fired, and the enemy—soon identified as German by his accent—falls to the ground. A series of (film) shots follow, crafting a ghostly picture of the corpse in close-up. Things seem clear at this point. The model person has killed the model enemy; task accomplished.

Act 2, beginning with a wipe two minutes into the film, is the psychological, educational, and narrative core of *Kill or Be Killed*. The film restarts. It begins with the first observational encounter between the two persons, sergeant and sniper. The viewer experiences the shifts in identification, self-presentation, and point of view that preoccupied the five hours leading up to the killing of the German enemy.

At the opening of Act 2, the viewer occupies the German's optical position and "sees through" the device of the concealed figure. She peers through his rifle sight. The screen and the sniper's telescopic sights fuse. The German voices: "Ja wohl—800 yards—wind, about a minute right." The viewer (both the film's viewer and the rifle scope's viewfinder) pans up until a distant figure comes into the center of the bull's-eye configuration. A rifle sounds, and the screen reverberates, marking the bullet's release. A cut to the British sergeant shows that he is unhurt; he wriggles out of the frame as water drips out of his shot-through canteen.

Back with the German, the viewer now seems to see through the German's binoculars. Or rather, perhaps she and the filmmaker are looking through binoculars from the German's vantage point. She is inside the body of the sniper (but apparently invisible); it is as if *her* hands are directing the horizontal leftward pan synchronized with the movement of the binoculars.

The effect of these shots, a "being there" both physical and emotional, is distinct from that of the standard point-of-view (POV) shot characteristic of the period in both fiction and nonfiction film. Lye's evocation of the experiential and corporeal can be best understood in terms of what media scholar Alexander Galloway calls "the subjective shot" and film theorist Edward Branigan calls the "perception shot." In Alexander Galloway's words, "the POV shot tends to hover abstractly in space at roughly the same diegetic location of a character. But the subjective shot very precisely positions itself inside the skull of that character."[46] Branigan dates the usage of the "perception shot" to Hitchcock's work at large as well as to a set of largely postwar narrative fiction films wherein "what is important is not so much that a character sees something, but that he experiences difficulty in seeing."[47] Although he lacks the telescopic sights of the Nazi, Smith, the British sergeant, struggles to anticipate the viewing position of the Nazi sniper. Whereas the German never audibly wonders if the Brit is watching him, the sergeant asks himself straight away: "He's probably watching the bridge. Gone, not a sign now." After the stalk, the sergeant is able to identify the sniper: "That's him. Binoculars flashing.... That's him. Aye that's him alright." In contrast, the Nazi is at a loss, sighting technologies notwithstanding: "Where is he. Might be anywhere." This is the point at which viewers recognize the model person's success, at least in terms of concealment from the model enemy. In the early section of the film, viewers were invited to look out through the German's sighting devices. Now, viewers see alternately through the sergeant's binoculars and from the position of the cameraman, himself concealed behind a tree and thereby introduced as a third party.

At the same time as Act 2's editing structure destabilizes the construct of a fixed point of view, its explicit narrative addresses how to camouflage oneself. Over the course of this section, the model person transforms himself from a position and appearance of vulnerability into one of proper concealment. Initially, the point of view moves between the abstracted filmmaker, the model enemy, and the enemy sighting devices. Only once the British sergeant has "dressed" in the materials of his natural environment, painting his face with mud and garnishing his helmet, can the viewer see within and through him, or at least through his eyes. One cannot see, unless one is already hidden—or so it would seem.

As he dresses, the model person blends in more and more with his wooded environment, also moving lower and lower to the ground. As the sergeant's thoughts are voiced:

> Net's O.K. for garnish if you've got the time. For a quick change, give me the band idea. A bit of dirt to tone it down. Easy to carry—just two clips and a band of string and rubber. That garnish in the net gets tangled sometimes. I'll be changing it pretty often today. Make these clips yourself. No waiting for an issue. That one's all right—this one's a bit tight. Fern. Birch. Aye, fern and birch. That'll do till I get to the hedges and meadows.

The viewer, at least *this* viewer, is reminded all along, however, that despite his best efforts, the sergeant is perfectly visible—perfectly visible to both the filmmaker and the viewer. Both have already found a second and invisible skin in the filmic layer itself. However here, as opposed to the farcical universe of Monty Python, there is no joke, either stated or implied.

Act 2 culminates in the "stalk-counterstalk" sequence. The camera follows the sergeant and the sniper as each simultaneously seeks the other. Hiding and seeking are not static activities conducted from fixed positions, as they are in "How Not to Be Seen." Rather, the stalk-counterstalk sequence involves resolving the simultaneous movement of the body through space and the immersion of the body *into* that space. As in Roger Caillois's description of the *camoufleur* who "breaks the boundary of his skin" and "tries to look at himself from any point whatever in space," one must move through an environment within which one is simultaneously integrating oneself.[48] Achieving this type of movement involves both self-fashioning in the style of the landscape and an internalized choreography.

In Act 3, the sergeant approaches the dead sniper and proceeds to remove the German's sniper suit, a disruptively painted textile smock. Throughout the film, the German has worn it to try to "blend into the woodwork," to disguise the shiny buttons of the

German uniform lurking beneath. Before his death, the German used the suit to disappear into the surroundings. In contrast, the British sergeant now uses it for purposes of mimicry—to play the part of a dummy lure. The sergeant takes the smock for his own later use. Removing the foliage garnish and propping the German's body against a post, the sergeant transforms the dead enemy sniper into a dummy live British soldier (figure 3.5).

Undressing the sniper hides him in a different sense; rather than being concealed by means of a camouflage suit, now he is a corpse dressed and posed so as to resemble something else. By sleight of dress, he has become a decoy. In this case, what a man is wearing underneath his disguise is simply another man's disguise.[49] Now visible, standing out in the field, the dummy attracts a whole group of uncamouflaged German soldiers at whom the sergeant immediately takes aim. The sergeant remains concealed as the enemies drop dead, one by one. The filmmaker and the viewer together are left positioned in the middle of an open field as, in the final shot, a fleet of reconnaissance and bomber airplanes fly by overhead. The filmmaker has crafted his *own* disappearance, above all.

The lack of voice-over narration and its replacement by competing internal monologues is a central device in the film (figure 3.6). And its value is not simply in adding to the drama of the story, but also, and crucially, in guiding the viewer through the various stages of awareness, observation, and identification required for a deep knowledge of camouflage. As a *Documentary News Letter* reviewer described: "There is no commentary, but the thoughts of both hunter and hunted are spoken, one by a Scots voice, the other in English with a German intonation, and this device adds enormously to the tension."[50] There is a structured eradication of direct identification with a single human subject or behavior. The subjectivity at stake in camouflage destabilizes the self, even as that self is shifted closer and closer to the British position. Meanwhile, the Nazi sniper repeatedly questions whether he is or is not able to see his British foe through his sighting devices and whether he will or will not be able to shoot at him effectively. As point-of-view shots alternate between the

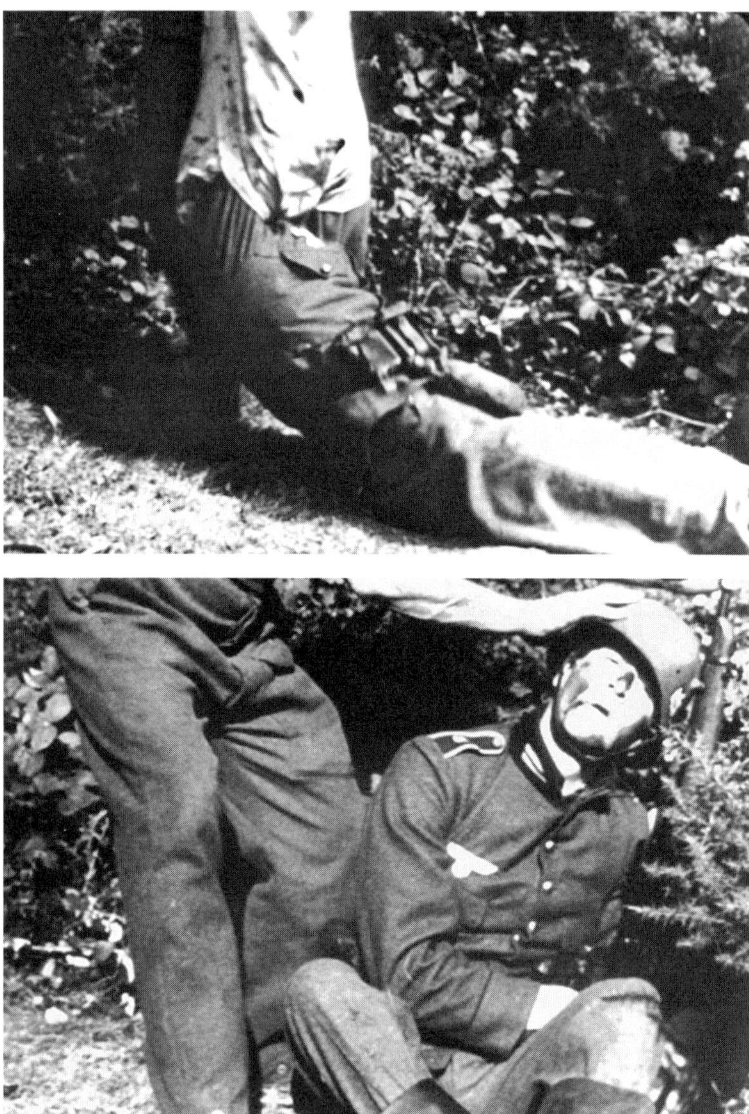

Figure 3.5—Making a decoy. The sergeant removes the dead sniper's camouflage smock and replaces it with a British military jacket, hoping to lure other Nazis to the scene. Reprinted courtesy of the Len Lye Foundation, Govett-Brewster Art Gallery.

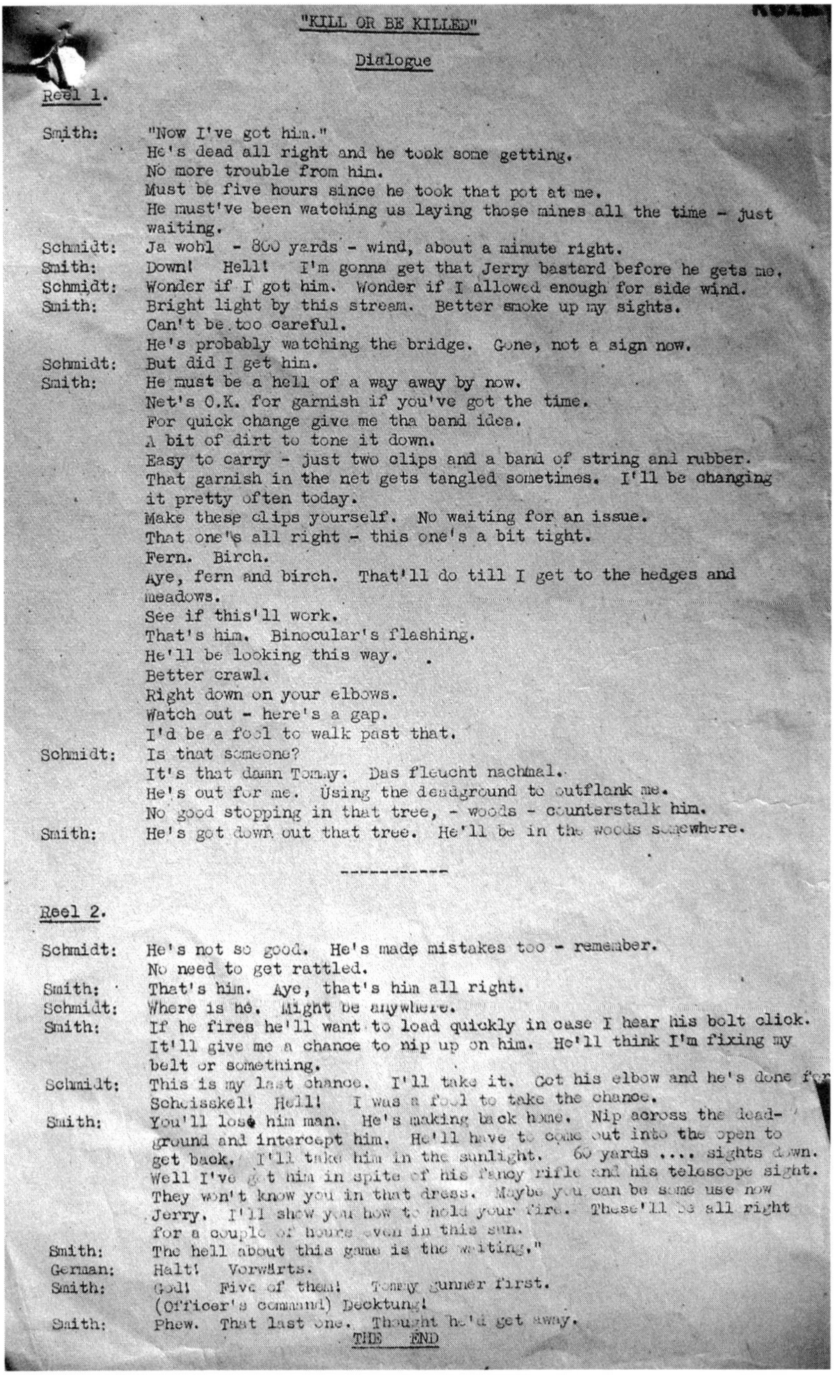

"KILL OR BE KILLED"

Dialogue

Reel 1.

Smith: "Now I've got him."
He's dead all right and he took some getting.
No more trouble from him.
Must be five hours since he took that pot at me.
He must've been watching us laying those mines all the time – just waiting.

Schmidt: Ja wohl – 800 yards – wind, about a minute right.

Smith: Down! Hell! I'm gonna get that Jerry bastard before he gets me.

Schmidt: Wonder if I got him. Wonder if I allowed enough for side wind.

Smith: Bright light by this stream. Better smoke up my sights.
Can't be too careful.
He's probably watching the bridge. Gone, not a sign now.

Schmidt: But did I get him.

Smith: He must be a hell of a way away by now.
Net's O.K. for garnish if you've got the time.
For quick change give me tha band idea.
A bit of dirt to tone it down.
Easy to carry – just two clips and a band of string and rubber.
That garnish in the net gets tangled sometimes. I'll be changing it pretty often today.
Make these clips yourself. No waiting for an issue.
That one's all right – this one's a bit tight.
Fern. Birch.
Aye, fern and birch. That'll do till I get to the hedges and meadows.
See if this'll work.
That's him. Binocular's flashing.
He'll be looking this way. .
Better crawl.
Right down on your elbows.
Watch out – here's a gap.
I'd be a fool to walk past that.

Schmidt: Is that someone?
It's that damn Tommy. Das fleucht nachmal.
He's out for me. Using the deadground to outflank me.
No good stopping in that tree, – woods – counterstalk him.

Smith: He's got down out that tree. He'll be in the woods somewhere.

Reel 2.

Schmidt: He's not so good. He's made mistakes too – remember.
No need to get rattled.

Smith: That's him. Aye, that's him all right.

Schmidt: Where is he. Might be anywhere.

Smith: If he fires he'll want to load quickly in case I hear his bolt click.
It'll give me a chance to nip up on him. He'll think I'm fixing my belt or something.

Schmidt: This is my last chance. I'll take it. Got his elbow and he's done for
Scheisskel! Hell! I was a fool to take the chance.

Smith: You'll lose him man. He's making back home. Nip across the deadground and intercept him. He'll have to come out into the open to get back. I'll take him in the sunlight. 60 yards .... sights down.
Well I've got him in spite of his fancy rifle and his telescope sight.
They won't know you in that dress. Maybe you can be some use now Jerry. I'll show you how to hold your fire. These'll be all right for a couple of hours even in this sun.

Smith: The hell about this game is the waiting."

German: Halt! Vorwärts.

Smith: God! Five of them! Tommy gunner first.
(Officer's command) Deckung!

Smith: Phew. That last one. Thought he'd get away.

THE END

Figure 3.6—Internal monologues. The "dialogue" in *Kill or Be Killed*, shown here in an official transcript, is in fact not a two-way conversation or set of verbal interactions. It is instead two internal monologues voiced aloud. The Ministry of Information included this transcript with a package of publicity materials distributed along with the film starting in 1943. [PRO INF 6/479].

perspectives of "model person" and "model enemy," the thoughts of both are voiced aloud, and the "dialogue" takes the form of a conversation between two characters who cannot hear each other. Only the viewer is in a position to complete and comprehend, through the construct of the film, the totality of events. This back-and-forth mode of narration again implicates the viewer in the position of the filmmaker. An oscillation between the voices, internalized by the viewer, activates what may be experienced as a sense of assimilation into the film itself. Although from time to time this may be broken up, as Manvell noted, by occasional titters.

Examining the structure of *Kill or Be Killed* in relation to that of a more conventional civilian training film ostensibly covering the same subject is instructive in teasing out the implications of the logic of camouflage for filmic pedagogy and self-concealment activity. *Camouflage and Fieldcraft*, produced by the Army Kinematograph Society (AKS), reproduces a model of military training based on the close-order (or parade-ground) drill, still the basis of British infantry training at the outset of World War II.[51]

Direct imitation and uniformity of appearance are crucial to parade-drill training and as such provide the narrative structure and pedagogical model for the AKS film. In parade-drill training, an individual learned to mimic exactly the behavior of the instructor—or drill sergeant. Likewise, in the AKS film, the viewer is requested to copy exactly the specific actions and bodily movements and appearances of a model person. As the opening statement implores: "The detailed movements and care of arms in this film should be carefully studied. Learn them—Practise them" (figure 3.7). The self formed by such pedagogical approach is that of the mimic or double—a replicant of the model self.

The off-screen AKS narrator plays the role of drill sergeant, and the model person acts as the narrator's puppet, a literal and highly visible illustration of the step-by-step lesson at hand, performing each described action (ghost walk, tiger crawl, and so on) in clear view of the camera and the viewer. The narrator, meanwhile, directly addresses both the viewer (mimic) and the model person.

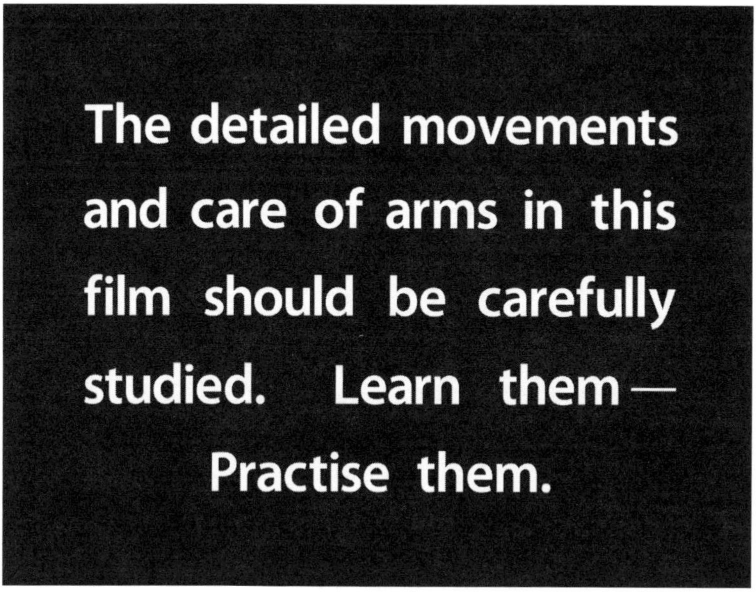

Figure 3.7—How to. The training film *Camouflage and Fieldcraft* (1942), produced by the Army Kinematograph Society, presented mastery of dynamic camouflage as a series of repeatable, memorizable movements. Graphic based on a still from the film.

He addresses the model person directly—as when he says "that's right," in response to correct behavior and obedience. At other points, the narrator draws attention to the distinction between the model person and the viewer, as when the narrator says "Don't carry it like this, with the muzzle leading," in apparent response to the model person's incorrect gun handling.[52]

In contrast, Lye cast the voices and bodies of his film's characters so as to place the viewer in precisely the position of someone who knows how not to be seen. Internal production notes emphasize a series of decisions Lye and his colleagues made regarding the casting of the project. As was prominently announced in subsequent publicity documents, "The part of the English sergeant is played by Staff Sergeant Duncan Chisholm. He knows how this job should be done, for in civil life, he is head *ghillie* at a large game park." The ghillie does not speak, however, but merely acts and moves through space. The viewer is privy to both his and the enemy's internal thought processes. Thus, *Kill or Be Killed* is structured in such a way as to narrate a way out of a doomed pedagogical model. Being able to mimic the actions or memorize the words of others would have done the sergeant no good. Rather, he had to be able to think for himself and improvise on the spot. Dynamic camouflage, as enacted here, is founded on principles of virtual projection into the position of the filmed surveyor, on the one hand, and practices of mixed-media collage, on the other. These mixed-media practices pertain in the first instance to activities of the subject in the protofilmic world. In the second instance, they pertain as well to the principles by means of which filmic instruction itself is constructed. The two instances are inseparable components of the subjectivity demarcated by dynamic camouflage. Effective camouflage, *Kill or Be Killed* teaches, comes from within, as well as from without. As a result, what is within and what is without merge, and the self disappears from the view of others.

Epidermal Cinema

The sergeant's behavior during Act 2 models the process of camouflage through the dressing of his body: its garnishing with leaves, mud, and branches. He begins to move in a fashion suited to self-concealment. He begins to move *with* his environment, rather than simply through it. Meanwhile, the German changes his tune as he recognizes his own position of danger. "He's out for me. Using the background to outflank me. Counterstalk him. Where is he? Might be anywhere." Whereas in the film's early section, the German's sighting devices dominated the viewer's point of view, now she sees alternately from the sergeant's perspective and the position of the cameraman, concealed behind a tree.

The camera's presence and movement both within and between points of the environment, as distinct from those of the "model person" or "model enemy," is manifest in the stalk-counterstalk sequence. Two long shots in the film actually take the viewer inside strategic concealment and are expressive of camouflage in a phenomenological sense (figure 3.8). The first such shot occurs as the

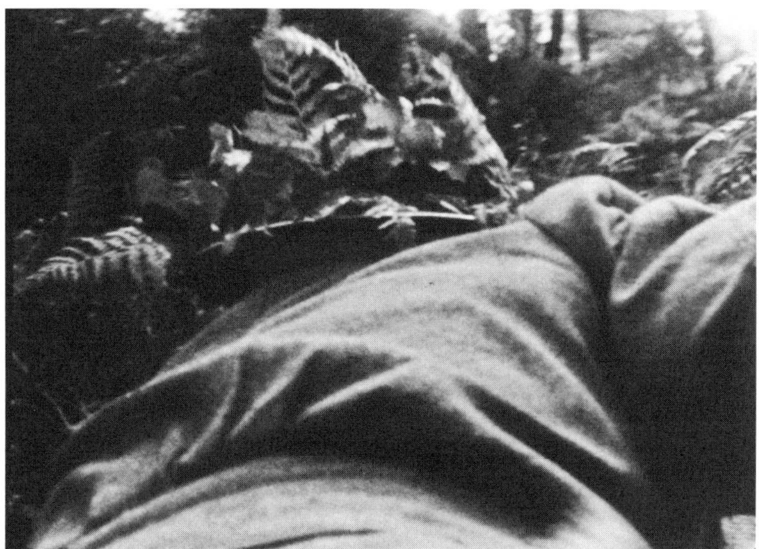

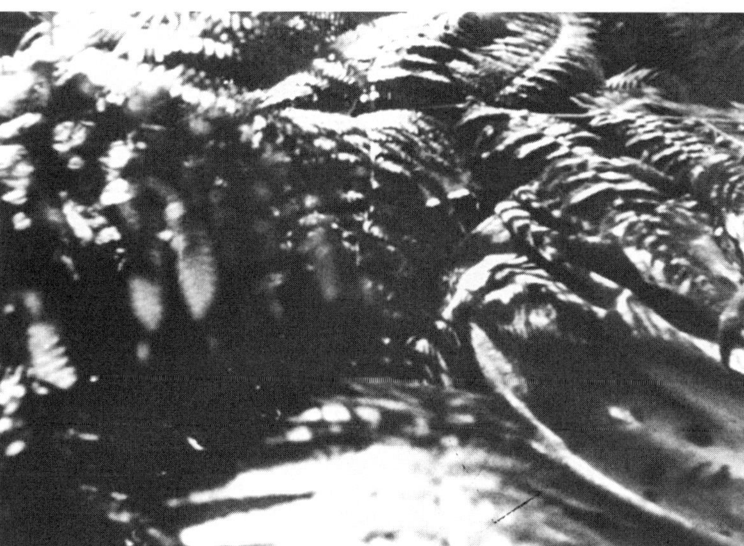

Figure 3.8—Epidermal cinema. The camera moves over and through bodies and underbrush with alarming fluidity. Reprinted courtesy of the Len Lye Foundation, Govett-Brewster Art Gallery.

model person prepares his halo of fern fronds and birch branches. The outfit change marks the point of transition after which viewers begin to see from his point of view, rather than through the enemy's reconnaissance apparatus. This shot also marks the point of entry for the film's only overtly didactic interlude: a recipe for how to make a successful corporeal garnish and modify it to suit one's specific environment.

The second shot occurs during the stalk-counterstalk sequence. The camera glides over and through the bodies and the underbrush. The filmmaker/viewer moves between positions and through space with an almost alarming fluidity. The camera navigates alongside both sides of the stalk, involved in its own honing practices, all the while unseen. It seems to enter the landscape itself, as if caressing the leaves at the same moment as the abstract patterns are registered as light on emulsion. "Pure cinema" here emerges as a perfected performance of hide and/as seek.[53] At these moments, in what one may call "epidermal" shots, the film itself serves as the viewer's own disguise.

These shots are expressive of this camouflaged subject position. The similarity between the epidermal shots in *Kill or Be Killed* and the textilelike direct cinema images Lye had produced earlier in his career were remarked upon by Alberto Cavalcanti just after the war: "This is pure documentary.... The patterns of the bracken as the camera crawls in the woods could have been designed by Len Lye himself in any of his earlier [animated] films."[54] Once understood as kindred cinematic designs, the animated and live-action images emerge as experimental media for the figuration of Lye called "pure motion," Cavalcanti called "pure documentary," and the *New York Times* called "pure cinema."

The "bracken" shots, to use Cavalcanti's term, the shots I called "epidermal," can be understood as the culmination of an alternative camouflage narrative running throughout *Kill or Be Killed*. This is direct animation that is also live action. Over the course of the film's central fifteen minutes, viewers encounter motion in an evolving relationship with the filmic frame. Motion progresses from being something operating *within* the frame (as in the opening shots, wherein viewers' attention is called to the wriggling sergeant in the lower left of the frame) to movement *as* the frame in the stalk-counterstalk sequence. The film plunges the viewer into the dynamic space of the filmic environment—to the "bracken," a medium for the concealment of its own materiality, as well as for the concealment of the human body wrapped within. To create this epidermal effect, Lye and his cameraman Adrian Jeakins hung their shooting camera, cradled by a leather belt, from a rifle not then in use as a prop. Standing side by side, holding the rifle between them, they turned on the camera and ran after "Schmidt," as he wriggled through the muck.[55] As the light hit the lens, the camera enacted its own ghillie crawl, becoming the real sniper in the scene.

Advice contained in the 1943 edition of *Sniping* pertains to film and field alike: "Vanish—but be there and see what is to be seen. You have all the earth to hide you, if only you call on her by using your wits, your eyes, your limbs."[56] Here, the viewer, operating through the camera, becomes a sniper in the film. Meanwhile, despite what seems to be his best efforts, the sergeant remains perfectly visible to the cameraman and the viewer alike. The instructional film that culminates in the epidermal shots is certainly not instruction in the performance of precise behaviors that directly imitate a model who loads a rifle or does math problems. Rather, the instruction is in how to immerse oneself simultaneously into the film and into the environment. The result in both cases is a version of a film constructed in the process of viewing, a productive cut in which the viewing self is both inserted and rendered invisible.[57] Full self-consciousness becomes literal self-observation in the cinema of dynamic camouflage.

Camouflage Consciousness

The strategic manipulation of one's immediate environment for purposes of self-effacement is at the core of this way of being—at the core of camouflage consciousness, to appropriate a phrase first used by the French architect Jean Labatut in 1943, the year of *Kill or Be Killed*'s release. With the American entry into World War II,

Labatut, a professor at Princeton, began a series of lectures in which he reflected on camouflage at the level of strategy, tactics, and material. With over fourteen hundred students in the audience of his course "Camouflage Consciousness and Camouflage Discipline," first part of the Princeton physical fitness program and eventually integrated into a Compulsory Training in Camouflage Discipline Program for architecture students, Labatut emphasized the need for what he called "a consciousness, discipline, and finally an instinct for camouflage."[58] In his view, camouflage education had thus far restricted itself to a basic, and inadequate, level of instruction. It was generally flawed from the outset insofar as educators restricted discussion to specific objects and tools, without addressing the larger issues of strategic self-awareness and visual training. As he told his students: "Too many people still think camouflage includes only paint, sniper suits, nets and all kinds of special materials." Putting questions of materials and objects aside, Labatut's view was that protective concealment demands primarily the development of new habits of mind. "The sooner you develop a consciousness of the importance of camouflage and accustom yourself to the discipline of protective concealment, the better it will be for you and for others on your side. Such consciousness and such discipline can help you to remain physically fit and prevent you from becoming a casualty."[59]

Direct mimetic performance was not the answer, according to Labatut, for there was no specific set of skills, no preordained set of rules to follow: "Mentally trained people are capable today of following even the most difficult scientific research, but the same people are lost when they are faced with new means which force them to develop their inner nature. The reason is that most of them cannot rely on their feelings or their state of consciousness as they can rely on their mental training."[60]

To remedy this state of affairs, Labatut proposed an alternative: "mass training in camouflage consciousness and camouflage discipline." To his students, he emphasized that there was no one single skill they needed to learn, but the sooner they mastered a set of skills, the better off they would be. "A protective metal armour is not always available, but camouflage consciousness plus camouflage discipline are armour you can use at will. The knowledge and the nature of such armour may mean the difference between life and death," he maintained. By acquiring camouflage discipline, he told them, you are "not only protecting yourself and others on your side, but learning at the same time how to detect the action and intention of the enemy." One must acquire what the praying mantis has, what Labatut calls an "inherited camouflage discipline."

"We shall not have time to teach you all of the 'How-to-do-this' and 'What-not-to-do' but we will ask you to think, to feel and to act with understanding toward camouflage," Labatut began his first lecture for "Camouflage Consciousness and Camouflage Discipline."[61] *Kill or Be Killed* would have been a perfect supplement to Labatut's syllabus. The filmic instruction that culminated in the camouflage epidermal shots in *Kill or Be Killed* lies not in the performance of precise behaviors, but rather a technological system in which the self is implicated and through which the modification of the self is structured. That technological system comes into being through practices of collage and innovation. As part of this process, the self is inscribed and reinscribed, while being rendered absent, within the contours of a larger media ecology.

As Lye said of his early innovation in kinetic art making, "I've been reduced to painting on film, or scratching film, not that I want to, but because I want to deal with the control of three-dimensional motion." In *Kill or Be Killed*, Lye crafted an education and experiment in animated self-concealment. Such a process might well characterize the activities of filmmaker and *camoufleur* alike, above all in Lye's animated and documentary works. Just as there is assumed to be a separation between organism and environment, there is also often assumed to be a demarcation between figure and ground within the film frame and between the frames themselves. Just as the sniper's goal is to break down the visual boundary between himself and his immediate environment through various techniques of bodily coloration, motion, and clothing, Lye's goal was to use similar means to break down the boundaries between figure and ground and between frames in the filmic medium. Dynamic camouflage, effectively performed through practices of psychic projection and mixed-media collage, could collapse the boundary between two positionalities, of filmmaker (observer) and film subject (the observed).

Lye's work exemplifies a logic of camouflage translated into an art of dynamic immersion in the form of both experimental and documentary moving picture production. Lye's *Rainbow Dance*, completed for the General Post Office three years before the outbreak of World War II, prefigures his live-action animation of strategic concealment in *Kill or Be Killed*. To the rhythms of abstracted Cuban jazz music, a city dweller carrying an umbrella is transformed through animation into someone who blends in with a rural landscape. In his essay on this film, "Experiment in Color," Lye writes, "The rainbow changes him into a color silhouette and his city clothes into a hiker's outfit."[62] His film exhibits the chameleonic figure, although it does not elicit viewer entry into that figure per se (figure 3.9).

The dancing man enacts the role of visual chameleon on film. He wears a costume—a second skin colored on film—that allows for his chameleonic color changes in relation to the shifting colors of his woodland environment. Like a chameleon, the dancer's signature changes from hunter green, to apple green, to white specked with yellow, and so on throughout the visible spectrum. But the mode of color change—of what appeared to be the man's chameleonic performance for the camera—was the artist-technician's own painstaking application of filters, gels, and paint strokes. In the nonfilmic world, where there was no animator to maintain such "colour continuities and chromatics," in Lye's words, even the best-made sniper suit could not match this performance. Only through the activities of the hands, mind, and body of its wearer—only through "camouflage consciousness and camouflage discipline"—could it help effect concealment in multiple environments. "Even if they have spotted you," as Idriess's *Sniping* informed, "you would have wriggled from there before they arrive."[63] Likewise, through animated and live-action film—via experimental and documentary representation—the environment

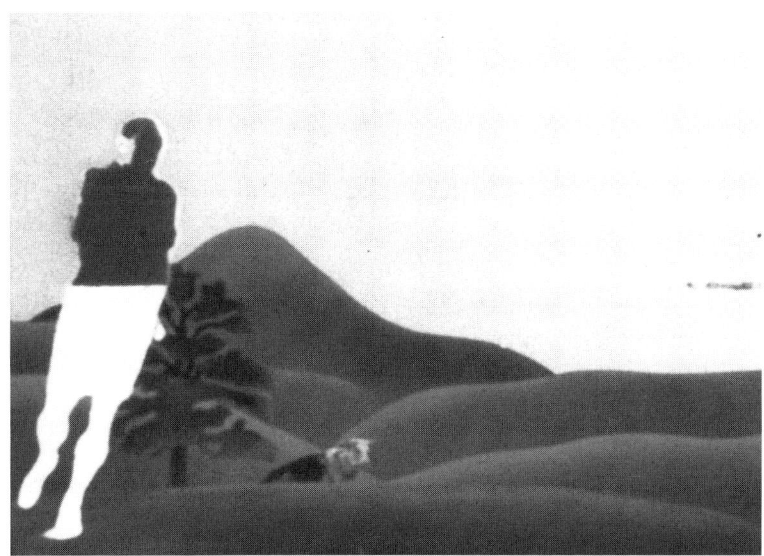

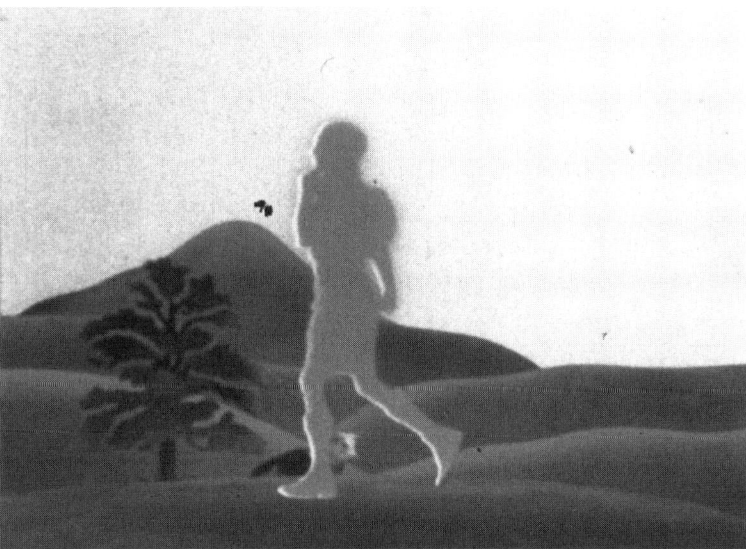

Figure 3.9—Rainbow dance. Lye's *Rainbow Dance* (1936) depicts a city dweller transformed through animation into a figure chromatically in harmony with a rural landscape. Stills reprinted courtesy of the Len Lye Foundation, Govett-Brewster Art Gallery.

could become a habitat and a resource base in which both to film and to embed one's own bodily outline in dynamic movement.

Film theorists have taken various approaches to articulating the relationships between viewer, camera, screen, and filmed object in connection with the experience of film. These accounts have been framed in phenomenological, analytic, and structuralist terms. Feminist theorists have drawn on Jacques Lacan's analysis of "screen," "look," and "gaze." For Lacan, in *Four Fundamental Concepts of Psychoanalysis*, seeing and being seen (the subject of the "gaze") are coconstitutive structuring concepts of psychic development. This psychic development is framed in both individual and social terms. In Kaja Silverman's formulation, "To 'be' is in effect to 'be seen'," so that "even as we look, we are in the 'picture' and so, a 'subject of representation.'"[64]

But not so in *Kill or Be Killed*. In *Kill or Be Killed*, the self is constructed as an object to be transformed, but as one "not to be seen," rather than one on which to gaze. Both *Kill or Be Killed* and Lye's animated projects invite the question of whether the connection between seeing and being seen is always valid. In *Kill or Be Killed*, after all, only by practicing "how not to be seen" can one (male or female, viewer or filmmaker) properly survive in the world. Seeing as not being seen corresponds to a form of subjectivity that involves a direct relation with and integration into the ecology of one's natural, social, and media environment.

Here, what I call "the logic of camouflage" corresponds with identification with an absent or hidden filmmaker figure, as opposed to spectator identification with the camera view itself. Vision is constituted by the cultivation of invisibility. Vision is constituted through identification with the filmmaker figure. *Kill or Be Killed* is structured in such a way as to narrate the movement of subject status from the conventional figure to a heightened state of filmic being—of being *in* and *as* the film. The camera, in turn, recedes.

The filmmaker figure is hidden from the viewer, and—for all practical purposes—concealed from the model person. The model person, it might be said, acts as if he does not see the filmmaker.

Alternatively, he actually is no longer aware of the filmmaker or is playing along and interacting with the filmmaker unbeknownst to viewers. Identification between viewer position and filmmaker position coincides with the development of skills to subvert detection by the camera. The subject as object is constructed as an entity to be kept hidden. It is immersed in the environment in such a way as not to be "shot"—captured by neither the enemy combatant nor the optical surveillance of the camera.[65]

The logic of camouflage evident in *Kill or Be Killed* is the translation into a structured and ordered behavior, into a cultivated form of subjectivity, of "a seeing into the world" that is activated precisely by the rendering invisible of the self. In his 1935 essay "Mimicry and Legendary Psychasthenia," in which sociologist Roger Caillois characterized as "psychasthenic"—psychotic—the individual who "breaks the boundary of his skin and occupies the other side of his senses," he explained the psychosis thus: "He tries to look at himself from any point whatever in space. He feels himself becoming space, dark space where things cannot be put."[66] But for Lye, this is not a psychotic condition, but rather an aesthetic and behavioral aspiration; it is camouflage consciousness. Such a process might well characterize the activities of filmmaker, sniper, and *camoufleur* alike. Embedding the self into the environment involves a process of psychic projection that is analogous to viewer attachment to the habitat of the film environment.[67] As Ion Idriess summarized camouflage consciousness in *Sniping*:

> Use the country; become part of the earth upon which you walk or lie or hide; make yourself invisible with leaves, or earth stains, or with lightly teased strips of bark, with the broad leaves of the jungle or the grass of the forest, with the rushes of the stream or the spinifex of the desert, with the wheat of the field or the seaweed of the seashore … with the charcoal of night or the ochres of the coloured lands, with the bracken of the creek or with tea tree bark to make you grey as the granite rocks. Use your wits and eyes to make you one with the very earth upon which you walk or hide. Nature places the very materials to hand, no matter where you may be.[68]

Kill or Be Killed, considered in the historical, aesthetic, and practical context of Lye's earlier work and ambitions as experimental filmmaker and mixed-media artist, provides a window into the nexus of fabrication and perpetuation of dynamic camouflage. An artifact from a seminal moment in the shaping of contemporary experiences of realism training, *Kill or Be Killed* reconfigured a standard model of educational filmmaking along terms consistent with the aspirations of camouflage and camouflage consciousness itself—enacted in both field and cutting room. "Discovery [in film] is something that is a workout on imagery I haven't seen done before," Lye reported.[69] As he further reflected, "I like to feel nature through its movement more than its shapes and colors and sounds, you know, sieved through my sinews, as well as my eyes and ears."[70] *Kill or Be Killed* turns this concept into a logic of media production. Entering into the film environment becomes a survival strategy for avoiding self-exposure and for maximizing one's own capacity for counterreconnaissance. Here, the media environment itself becomes a site for productive resistance by the viewer turned *camoufleur*. The result in *Kill or Be Killed* was a form of filmed experience or "true story" that might properly be called a simulation. Here, a mode of filmmaking emerges that is aligned with a military pedagogy itself bound to principles of immersion and realism, rather than to the direct mimicry characteristic of standard military training that previously prevailed both in the field and in film.

First-Person Shooter

Lye's audio-visual training in *Kill or Be Killed* involved the destabilization of certainties—the *whats*, *whens*, *wheres*, and *hows* of model, enemy, and environment. And it was precisely this destabilization that appealed to a multiplicity of audiences. *Kill or Be Killed* is "well worth your while to hunt it out—that is, if you don't mind being shaken up, frightened and fascinated,"[71] wrote one reviewer in 1943. It gripped its viewers in ways that resonate strongly with the modes of interactivity, narrative enlistment, and subjective perspective that surround us today.

What was so striking about *Kill or Be Killed* upon its release—what made it stand out from standard infantry training of its era—was an innovative aesthetic and narrative structure perfectly suited to a regime of dynamic camouflage. The logic and aesthetic of dynamic camouflage inscribed in *Kill or Be Killed* (and notably absent from *Camouflage and Fieldcraft*) have, in the years since World War II, become increasingly pervasive in the media products of the military-entertainment complex, from action movies, to video gaming, to Virtual Reality (VR)-based combat training and battlefield trauma recovery.[72]

In the Cold War action movies of the 1980s, for example in the 1987 John Tiernan film *Predator* (original name *Hunter*), and especially in the increasingly immersive video game and military combat simulation experiences of the 1990s and 2000s, media manipulation in the service of self-erasure is the rule of the (war) game. In the twenty-first century, first person shooter (FPS) games—including such popular stealth FPS systems as *Metal Gear Solid 3: Snake Eater* and *Call of Duty 4*, and military recruiting games such as *America's Army* and *Battlefield*—provide entertainment and infantry training to millions of gamers.[73] Producers of these games have formed partnerships with the U.S. Army and Marine Corps, among other military operations. Versions of the games are increasingly used as tools in basic training and are considered to be valuable war-simulation products, tools for the cultivation of a twenty-first-century "camouflage consciousness."

Dynamic camouflage is implicit in the aesthetic, narrative, and play structure that forms the basis of these FPS games. They adopt a first-person perspective that presents the game environment from the perspective of the player character—embodied and therefore immersed psychologically as well as physically in the digital landscape. The fundamental activity of these games—its lifeline—is strategic concealment within the digital ecology of the game environment, this being a prerequisite to the violence that often seems to be the games' defining feature.[74] Some, such as the popular *Metal Gear Solid 3: Snake Eater* and *Call of Duty 4*, as well as the animal themed

Big Game Hunter, incorporate camouflage at both a strategic and a logistical level. In the case of *Metal Gear Solid 3*, a menu makes available various face paint and textile-patterning schemes for selection by the player. As she navigates the shifting visual contours of the digital environment, a camouflage index readout constantly updates her on the success (measured in percentage points) of her stealth methods at any given point. As the game proceeds and both narrative elements and environmental conditions evolve, the player's concealment techniques must shift accordingly. Original camouflage options provided from a menu become less and less effective; new camouflage options lie hidden, however, to be discovered by the keen sense of the player inhabiting the game environment. Further into the game, however, none of the revealed premade camouflage patterns are effective. The player must begin to layer one atop another, combining them in increasingly complex ways. Strategic concealment is enacted through cunning deployment of the specific items found in one's environment—leaves, dirt and modified textile patterns—as well as through the creative use of mouse and hand controls to simulate seamless humanoidlike stealth movements within the game.

On one popular Web site devoted to perfecting one's engagement in the first-person shooter environment, the player is instructed:

> If you can learn to dance, you will live longer.... Always try to keep your back against a wall, hill or object, it's one less place the bullets can come from. Make sure to look over your shoulder frequently, there's always someone covering your back, make sure it's a friend. Never move on top of a hill, you create a good silhouette target. If you learn the basics of movement, then you live longer, and can learn more.[75]

Or as the World War II sniper manual put it, one needs to "learn the capabilities of your extremely adaptable body—its speed, its immobility, its vanishing tricks."[76]

A chief goal for the FPS player, then, is to learn to move to the rhythm of camouflage—in a "rainbow dance" of pixels—matching one's own hue, shadow, texture, and patterns of movement with the frame rate and resolution of the media environment.[77] The supreme position is one of total effacement, of nonrepresentability, or at least of apparent absence. *Kill or Be Killed*'s articulation of the genealogy of dynamic camouflage provides a means of understanding the logic at work today in a range of twenty-first-century multimedia environments in terms of both their production and the experience of their inhabitants, those who choose or are required to occupy positions therein.

In addition to FPS gamers and infantry in training, that population also includes media hackers, who in addition to acting as players or subjects in the "ecology" of the FPS games also are producing or conceiving "mods" or "machinima," both forms of emergent gameplay, the former referring to new interactive forms and the latter to narrative films, both crafted using 3D real-time animation with video and other assets drawn from preexisting games. The process required to produce a work of machinima is not unlike the found-footage filmmaking of Lye's *Trade Tattoo*. Pieces of code—the basis for the real-time structures of the first-person shooter game environment—become the building blocks, the editing "shots," for a montage that becomes a singular story embedded in what is already a simulated world. An examination of these digital media environments in relation to the emergence of dynamic camouflage in World War II allows a fuller sense of the degree to which conceptualizations of gaming and interactivity are always already encoded with an ethos whose origins date to a moment in the convergence of natural and filmic worlds, a moment itself bound by material artifacts and the exigencies of embodied self-effacement.

In "How Not to Be Seen—a BF2 Machinima," by Danish gamer Gert Nielsen aka "[-A-] Washburn," first produced in 2005 John Cleese's voiceover from the Monty Python skit now describes a series of scenes from the FPS game *Battlefield 2* in which the people shot or blown up in the course of the original skit are shot or blown up in its representations of landscapes and buildings, using the weaponry resources of the game.[78] But in the machinima, framing it at the beginning and the end, the appropriation of the Monty Python skit is shown to be appearing on a large screen encountered in a jungle setting by two camouflage-clad *Battlefield 2* infantry troopers, who stand in metonymically for and mirror the subject position of the viewer actually viewing the mechanism's hack of both the game and the skit. Near the end of the machinima, after the simulated projection of the Monty Python original has come to a close, the *Battlefield 2* troops discuss the hacked skit they have watched, and by extension its appropriation and restaging currently underway. In voices appropriated from Sattler and Waldorf, the two old men from the Muppets who often play grumpy theater critics, the *Battlefield 2* characters now critique the project in which they are appearing in a crescendo that descends quickly from "That was wonderful" to "It was awful, boo!" The closing sequence lists [-A-] Washburn as director, producer, and cast, then a list of thank-yous and audio clips used. The video ends with a "copyright Gert Nielsen .a.k.a. [-A-] Washburn" accompanied by a superimposed ten-second extended screen grab of what is presumed to be his desktop editing station on which the "How Not to Be Seen: A BF2 Machinima" timeline is open and playback on the miniature viewer on the upper right of the screen is in progress. Monty Python's ironic appropriation of and commentary on the genre of the World War II training film thus gets appropriated, then supplies its own ironic commentary on both appropriations in a simulated environment in which simulation itself is taken to be the norm.

Monty Python's "How Not to Be Seen" works as an ironic commentary on an experience assumed to be already distant from its audience, an experience in which simulation was not the norm—something that had to be learned, worked at, achieved. Appropriation of the Monty Python skit by the machinima makers doubles the ironic appropriation of the training-film genre by Monty Python, in effect sampling that irony in an environment where simulation can both be taken for granted as the norm and therefore also be commented on ironically. In the context of this increasing self-reflexivity, camouflage consciousness comes to consciousness of itself, situated in a terrifyingly twenty-first-century mode of entertainment as warfare. This is a model of subjectivity in which the media maker, viewer, and inhabitant of the visual field are fused into one disembodied body, invisible to the media maker and without a camera in sight.

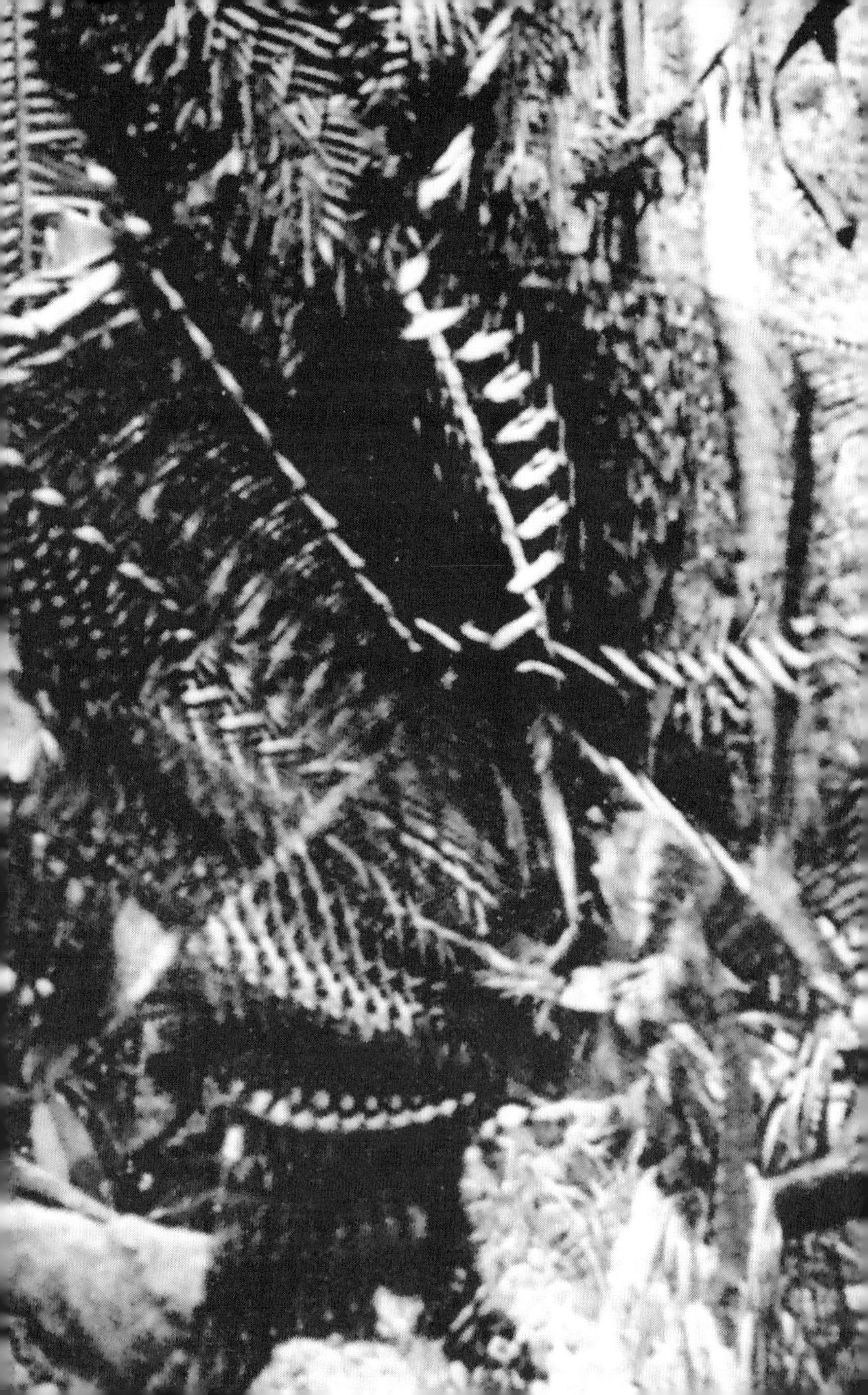

CHAPTER FOUR

Subject to Change

Skin as Screen

In his 1974 science-fiction novel *A Scanner Darkly,* writer Philip K. Dick envisioned a membrane that would render its human wearer impossible to identify. A quartz lens connected to a million and a half fractional representations of different people "projected outward in all directions equally," and "as the computer looped through its banks, it projected every conceivable eye color, hair color, shape and type of nose, formation of teeth, configuration of facial bone structure—the entire shroudlike membrane took on whatever physical characteristics were projected at any nanosecond, then switched to the next."[1] Dick called this invention a "scramble suit," and in the novel and in its innovative 2006 film adaptation, it is portrayed as an effective camouflage technology, what the modern military refers to as "active camouflage."[2]

In Richard Linklater's film version, the success of the scramble suit (donned by a digitally filmed and then rotoscope-animated actor Keanu Reeves playing the character Agent Fred/Bob Arctor) coincides with a corresponding erosion of Reeve's character's identity and indeed with an ascent into a state comparable to Roger Caillois's conception of "legendary psychasthenia." An undercover drug cop at home, and an undercover addict at work, Reeves's character uses the scramble suit to blend perfectly into a dynamic sea of human appearances and social identities. The more Agent Fred/ Bob Arctor seamlessly integrates into two sides of an environment constructed around visual technologies of police surveillance, the more he loses himself; cognitive dissonance and visual evanescence eventually spiral together into a state of psychotic dissociation. He indeed, as in Caillois's formulation, "feels himself becoming space, dark space where things cannot be put."[3]

The scramble suit made positive identification of its wearer impossible. The apparatus effectively eradicated the distinction between the visual identity of the human subject and that of an infinite number of other human bodies without any conscious effort on the part of its wearer. Translated into an apparatus for use in the world we all in fact inhabit (rather than an animated simulation) the scramble suit would seem to render the work of achieving camouflage consciousness automatic, though sometimes treacherous at a psychological level, a translation of camouflage discipline into the realm of reflexive instinct, albeit instinct aided by computers and live-action animation.[4] The suit would allow its wearer to change his or her colors. Which is to say, in brief, that it would allow a human wearer to act the chameleon.[5]

Over the course of the twentieth century, photographic camouflage has appeared in different forms (of which Linklater's may be seen as one cultural manifestation), responding to different modes of surveillance and employing different materials. All along and in every iteration, however, there has been an impulse, a discourse, an aspiration for a technology more encompassing, not one that would operate at the level of the landscape, trench, encampment or artillery store, but at the level of the individual human. This aspiration is in a sense for the skin itself to become a screen, for effective camouflage coloration to be seamlessly integrated with the human body, for dynamic camouflage, in essence, to be made a less demanding physical and psychological chore and more of an effortless compulsion—and one with no off switch.

How can we understand this impulse? From where did it emerge? Charting its history reframes the history of photographic camouflage and camouflage media that has been the subject of this book thus far. One way to investigate the chameleonic impulse is to walk through the historically and conceptually linked species of

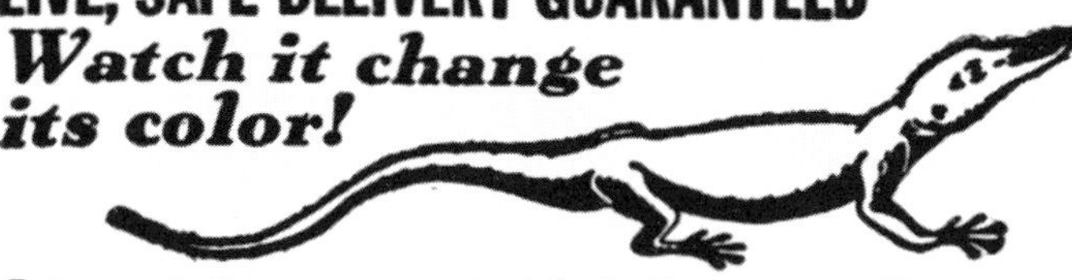

Figure 4.1—**Chameleons by mail**. A wide range of media and entertainment outlets (magazines and books, as well as museums and films) attempted to capitalize on widespread curiosity about, and desire to witness, real-time color change. *Science and Mechanics* (November 1934), p. 615.

camouflage one more time, this time using the chameleon as a guiding thread that allows us to rethink and resituate what we have seen so far (figure 4.1).

The Chameleonic Impulse

The chameleon's potential use as a model for technological innovations in strategic invisibility was noted about the same time as Abbott Thayer's early investigations into camouflage. The eleventh edition of the *Encyclopedia Britannica* (1910–1911), for instance, included a brand-new entry, "Colours of Animals." Its author, Edward Poulton, referred to the fish, amphibians, and reptiles capable of changing their colors ("the chameleon affording a well-known example") to any tint "which would be appropriate to a normal environment" as possessing "the most perfect cryptic powers."[6]

Later that year, in the summer of 1910, new exhibit labels were added to selected tanks at the New York Aquarium in lower Manhattan: "This species may change color at any moment."[7] The new label highlighted certain tropical fishes' capacity for chameleonic color change, the rapid transformations of their scaly skins' pigmentation in relation to changing environmental color conditions.

The added text served as both a warning and an advertisement. It admonished visitors to keep their eyes peeled; watch out, or both the fish and its color might disappear without your seeing what was happening. It also implied that what you, among the visitors *wanted* to see was the color change. But if you failed to look closely, you'd lose sight of what you were seeking, which is to say, both the object and its enacted transformation. In a promotional article published in the popular magazine *The Century*, aquarium director Charles Haskins Townsend explained that "changes of color occur hourly, may, in fact, occur at any moment, and are usually instantaneous." The change could happen right now or tomorrow morning. One could never know. Generally, however, the director reassured visitors that "a few minutes' observation is enough."[8]

Trying to depict this phenomenon in *The Century* article "Chameleons of the Sea" presented another dilemma. How could the color change of the chameleonic organism—the queen trigger fish, for example—be represented? The director commented on the difficulty of producing photographs that would depict the key characteristics of the chameleonic fish which he "placed in a small portable aquarium and carried out of doors in order to photograph their markings. This involved considerable handling of the specimens, and all the photographs secured showed only those phases of coloration and marking peculiar to frightened and hiding fishes."[9]

Might there be another way of modeling dynamic color change in a two-dimensional image to be circulated outside the aquarium or reptile hall? Townsend's and others' intense curiosity about dynamic color change in the natural world coincided with engagement with the logics and limitations of static and serial camouflage by both artists and militarists—as well as with the birth of the cinema, first as a "cinema of attractions"[10] and then by the 1910s and 1920s as a vital narrative form.[11] The three structural formations of camouflage have coexisted since the mid-twentieth century, and each contains aspects of the previous methods of photographic camouflage. Each is an individuated form of self-awareness that is also part of a network of institutional practices, and though they developed in sequence, they did not replace one another. The static, the serial, and the dynamic, previously treated largely independently, intertwine in the historical development of the chameleonic impulse.[12]

The Natural History of the Chameleon

In the late nineteenth century, evolutionary theorists, for all their discussion of animal coloration and adaptive concealment, said little, if anything, about what would seem to be the most striking manifestation of such protective coloration: rapid, that is, *chameleonic*, color change, as opposed to the static phenomena of mimicry or general protective coloration.[13]

Natural historians as far back as Aristotle and Theophrastus had been fascinated by these color changes. For natural philosopher Erasmus Darwin (Charles Darwin's grandfather), writing in the late eighteenth century, the myth of the chameleon provided a prime example of animals' "curious" ability, as he understood the phenomenon, to draw the color of their environment into their skins and, over time, actually to pass the absorbed colors onto their progenitors. In *Zoonomia*, he speculated that, "like the fable of the chameleon, all animals may possess a tendency to be colored somewhat like the colours they most frequently inspect."[14] Later, Erasmus's grandson Charles noted his sightings of chameleonic creatures during his voyage on the *Beagle*. In 1832, Charles Darwin wrote from Rio de Janeiro to his mentor, J. S. Henslow, regarding "an Octopus, which possessed a most marvellous power of changing its colours…evidently accommodating the change to the colour of the ground which it passed over—yellowish green, dark brown & red were the prevailing colours; this fact appears to be new, as far as I can find out."[15] Such beguiling observations or discussions thereof never made it into his later formulations of the theory of evolution, however—neither the original text of *The Origin of Species* (1859) nor the revisions that followed.

Chameleonic color change confounded nineteenth-century evolutionary biologists, who tended to think in terms of geological and

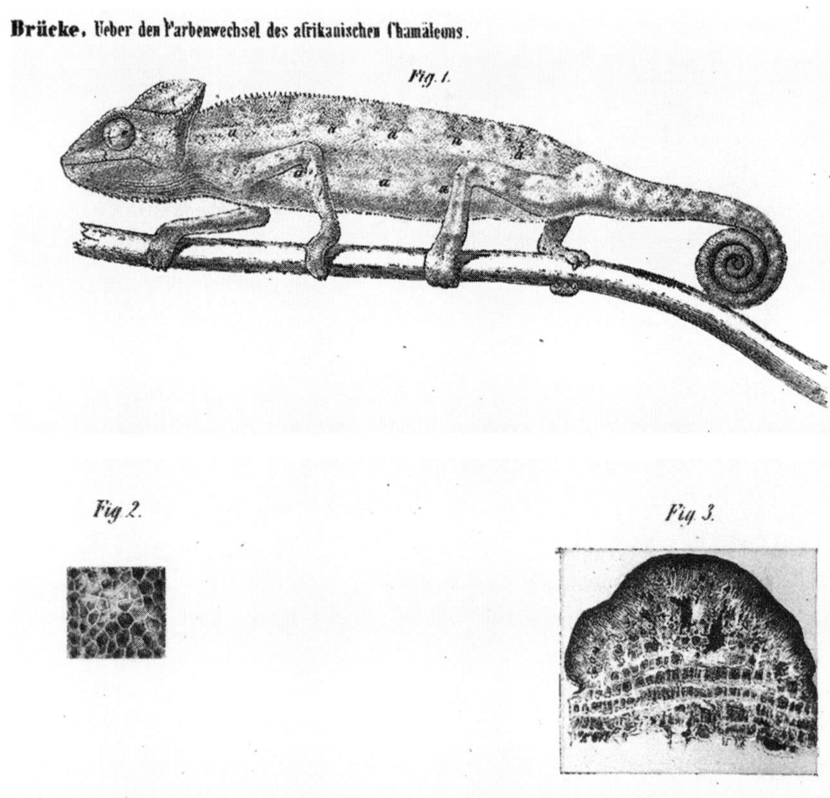

Figure 4.2—Chromatophores. Reproduced from Ernst Wilhelm von Brücke's *Untersuchungen über den Farbenwechsel des afrikanischen Chamäleons* (Vienna, K. K. Hof-und Staatsdruckerei, 1852), p. 35.

generational time scales, periods of hundreds, if not thousands of years. As Poulton stressed, chameleonic concealment was nothing if not fast: "Rapid adjustable protective resemblance is a response to dynamic conditions, in which, owing to the rapid movements of the organism, one environment is exchanged for another with speed, and any number of times."[16] Although naturalists Alfred Russell Wallace and Henry Walter Bates, as well as Charles Darwin, promoted general protective resemblance as exemplary of adaptation by natural selection, they seemed unable to make heads or tails of color-changing cephalopods, among other real-time color changers. In the chapter titled "On Natural Selection" in *The Origin of Species*, Darwin evoked leaf-eating insects, mottled-gray bark eaters, and "red grouse the colour of heather" and acknowledged that these schemes must play a role "in preserving them from danger," but made no mention of the rapidly changing forms of such coloration.[17] Physiologists elsewhere in Europe, however, were studying the mechanism of change—specifically, what in the chameleon or octopus skin causes the real-time color change (figure 4.2). The Italian G. Sangiovanni first identified the specific "color organs" in cephalopods, designating them *cromofori* (chromatophores) in 1819, and further study on their form and function continued throughout the century.[18]

In the absence of an evolutionary framework for exploring such a peculiar phenomenon, chameleonic capabilities remained examples of the older idea of "nature's disguises." Color change made evolutionary sense to these scientific inquirers only in relation to generational or seasonal time frames. In the famous generational case of the peppered moth, for example, as the Northern England landscape became covered with increasing amounts of industrial soot, the peppered moth population became darker and darker, blending in with the blackened leaves and trees on which the moths perched. The darker moths' success and the decline of the lighter variety of the animal were the result of the selective advantages of an environmentally synchronous palette. The arctic hare was a prime example of the seasonal case, changing from white in winter to brown in summer. In both cases, scientists could depict or illustrate these cryptic shifts in single or serial images or even taxidermy displays. The change from summer to winter could be tracked through two pendant paintings or by two specimens perched side by side, one in winter and one in summer dress.

Abbott Thayer was fixated largely on such static representations of the relationship between an organism and its environment. His focus was on the singular, perfected state of concealment in relation to a particular background, such as that captured in a photograph that seems to contain no animal at all.

When it came up, Thayer seems to have been puzzled about how to deal with the dynamic cases of chameleonic change, working as he did in a world of posed skins and canvas cutouts; he did not choose to mention them in any of his essays on protective coloration from the 1890s and early 1900s. In his book with his son Gerald, *Concealing Coloration in the Animal Kingdom*, the chameleon did make a short appearance in an appendix authored by Poulton, with whom Abbott had produced an exhibit of disappearing countershaded ducks for the Oxford Zoological Museum and had coauthored an article in *Nature* in 1902.

In this appendix—titled "Some Observations on Color Change among South African Chameleons"—Poulton presented "scattered notes" made in the company of another biologist and a physicist regarding what Poulton referred to as "the automatic adjustable counter grading of shadow on the two sides of the chameleon."[19] He and his colleagues had watched in surprise as an organism changed through the shifting conditions of the environment in which it had been placed and now maneuvered. Poulton compared the chameleonic color change to the "disappearing" duck exhibits he had previously cocurated with Thayer at two British museums. In the case of the chameleon, change was induced within the organism itself, not by a museumgoer turning a hand crank: "After it had been kept for some time in the dark it became of the brightest apple green. On exposure to light it darkened. Placed on a dark 'uniform-case' near the window in bright light it darkened along the dorsal area.... This appears to be a most interesting adaptation—a *dynamic* manifestation—of the principle discovered in its *static* form by Mr. Abbott H. Thayer."[20]

But here the animal took over what artists and scientists had, thus far, left off. As Poulton explained in *The Colours of Animals: Their Meaning and Use, Especially Considered in the Case of Insects* (1890), such "rapid changes of colour are due to changes in shape or position of superficial pigment cells controlled by the nervous system. That the latter is itself stimulated by light through the medium of the eye and optic nerve has been proved in many cases."[21] Unlike the rest of the volume, this section presented no visual illustration of its subject. What kind of adequate illustration might be possible within the confines of the printed page? It was, in fact, beyond the capacity of the Thayers' mode of visual representation.

Representing the Chameleon

How could one illustrate the acts of dynamic transformation Poulton had described? Manifesting dynamic color transformation in visual form seemed to be beyond the capabilities of a photograph. It was hard to fit in a single frame.

Returning to Townsend's dilemma when faced with illustrating his *Chameleons of the Sea* exhibit at the New York Aquarium: one could show one phase, and another phase, but the state and normal experience of the skin's transformation itself was impossible to photograph. If we inhabited Thayer's mental world, we would use a stencil to find out what the fish would look like were it hidden in the photograph. If the concealment were perfect, we would end up with the same problem we ran into in figure P.1 (see page 8): How do we know there is an organism, chameleonic or otherwise, there at all? One strategy might be to photograph individual color phases—to capture one moment and then the next. This posed problems both practical and conceptual, however, since the changes occurred precisely while the fish was in motion.

Alternatively, one could depict their multiple color transformations through something like the painting that Charles R. Knight made for *The Century* magazine (figure 4.3). Knight, a natural history painter known by 1910 for his dramatic murals of dinosaur

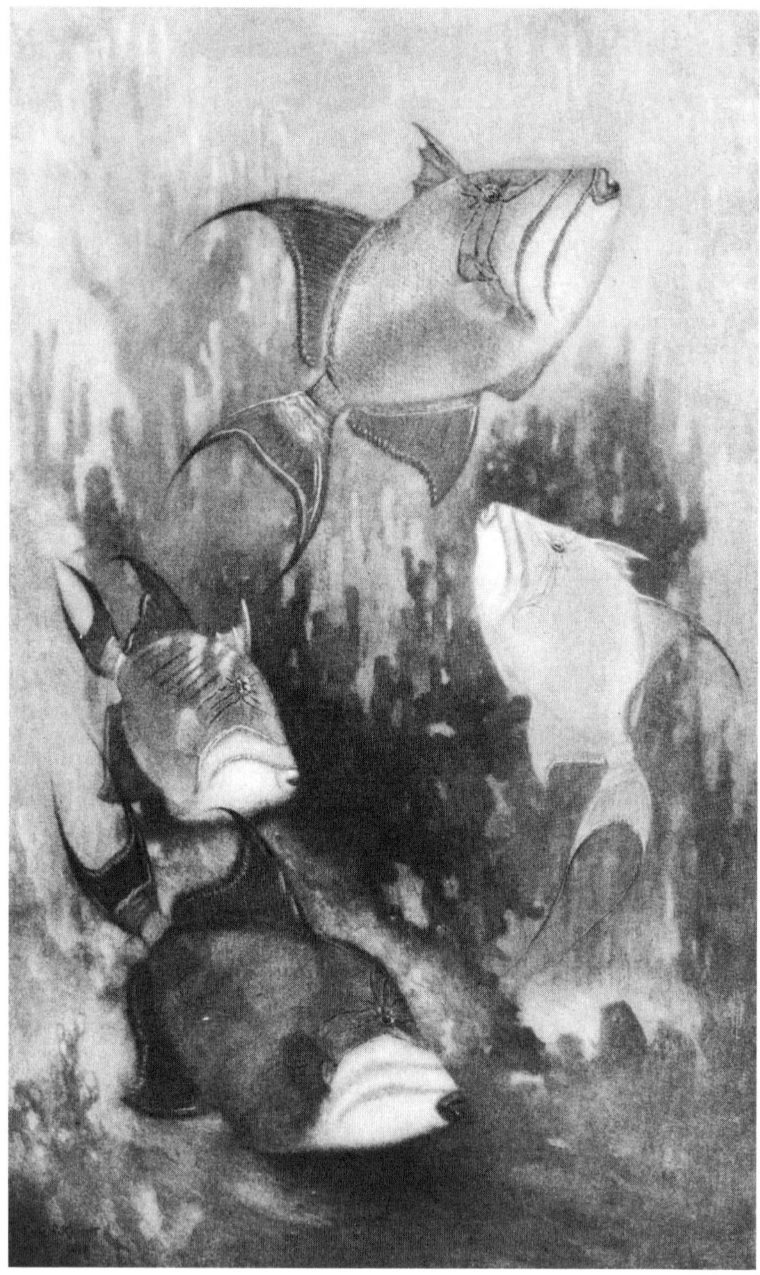

Figure 4.3—**Chameleons of the sea.** Charles R. Knight's painting of queen trigger fish illustrated Townsend's "Chameleons of the Sea: Some New Observations on Instantaneous Color Changes Among Fishes," *The Century* (September 1910), p. 1.

evolution and fossil formation at the American Museum of Natural History in New York, attempted to compress four different color phases of the queen trigger fish into a single painted frame. In the resulting serial portrait, a single fish is shown four times, in four different outfits, one might say. In some ways, the effect is similar to that of the type of time-based photographic series pioneered by Muybridge and Marey that were drawn upon in World War I reconnaissance. Indeed, physiologists interested in the mechanism of color change debated methods of sequential photographic documentation throughout the 1920s and 1930s (figure 4.4).[22] But here there is no statement of the interval between one phase and another. The sense of change over time is still absent.

Even when one could take a proper photograph—or series of photographs—it would not capture the specifically chameleonic and real-time adaptive aspect of the chameleonic color change. And the gap was of critical importance. Within this gap, the dynamic organism and the dynamic environment shift their contours and shadings enough to converge visually in a sort of color harmony—what Len Lye would later refer to as "color continuity" in relation to "progression and movement" in his 1940 essay "The Man Who Was Colorblind."[23] But could the gap be captured in a motion picture? Filmic representation would seem ideally suited, if anything could be, to documenting the essence of the trigger fish, the South African chameleon, and other creatures with color-adjustable skins. But as of the early 1900s, this would not have been possible with regular film—in the absence of true-color dye-coupled film processes, representing color change would have required delicate color painting or processing techniques such as those from which Lye and other experimental artists drew inspiration in the 1920s and 1930s.

That painted animated film had some relevance to chameleonic color change for the filmmaker in the early twentieth century is borne out through further examination of Lye's *Rainbow Dance*, completed in 1936, which combines the focus on kinesis and concealment that emerge in Lye's earlier direct animation with that later manifest in *Kill or Be Killed* (see figure 3.9). In the film,

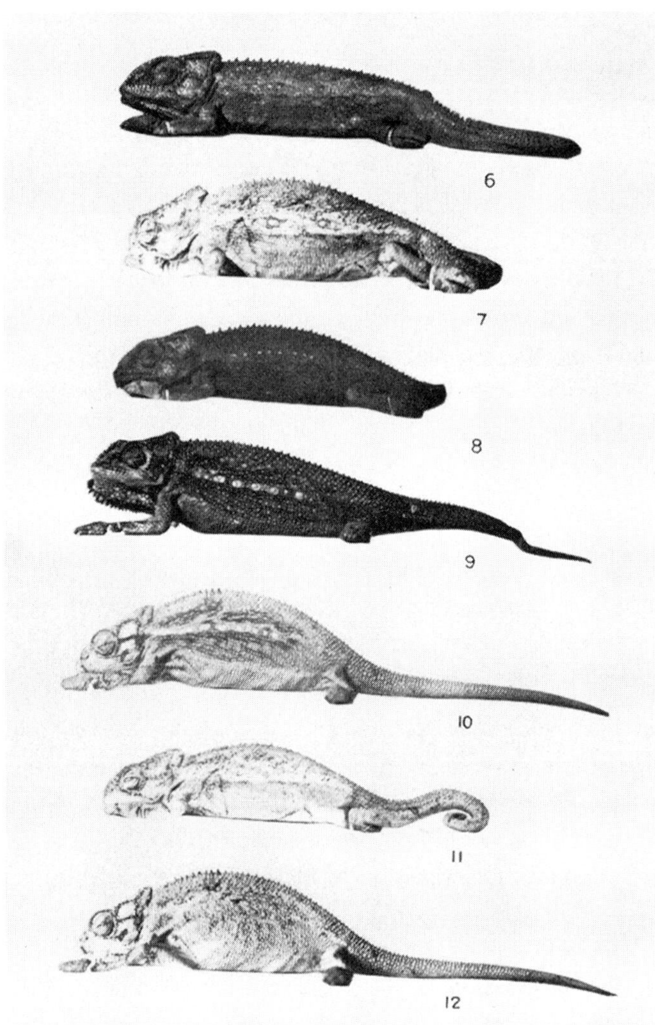

Figure 4.4—Photographs of dynamic color change. South African physiologist Alexander Zoond used cameras flashlight units, and a number of black-and-white film stocks to study rapid color change in the African chameleon and related species. Photographs served as key evidence in publications in the early to mid-1930s. Original caption for this image reads: "6: A chameleon on black cardboard in daylight; 7: The same animal on white cardboard in daylight; 8: The same animal on white cardboard in lamplight. Flashlight photograph; 9: A blind animal on white cardboard in daylight; 10: The same blind animal on white cardboard in lamplight. Flashlight photograph; 11: The same animal as in figs. 6–8 in darkness. Flashlight photograph; 12: The same blind animal as figs. 9 and 10 in darkness. Flashlight photograph." Reprinted from Alexander Zoond and Joyce Eyre, "Studies in Reptilian Colour Response: I. The Bionomics and Physiology of the Pigmentary Activity of the Chameleon," *Phil. Trans. Series B, Containing Papers of a Biological Character* 223 (1934), p. 54.

movement and adaptation are rhythmically linked, and animation is both the means and the catalyst for transformation. As Edward Poulton remarked in his 1910 essays on evolution: "Rapid adjustable protective resemblance is suited to wandering forms which must in the course of their lives continually pass and repass over environments of different colours."[24] Or as Lye wrote in his essay on the film, "Experiment in Color": "The rainbow changes him into a color silhouette and his city clothes into a hiker's outfit." To realize this experiment, Lye employed only black-and-white objects:

> No colour was used on the sets, where every object was painted in terms of black and white. For instance, a green hill…was painted in terms of black and white and photographed continuously for the red record.… A silhouette of a man was superimposed over each colour record in densities according to the dye required for colour.… Our colour would be clean and not suffer from any opacity of photographic colour light.[25]

Here, the dancing man plays the role of chameleon. He wears a costume—a second skin, colored on film—that allows for his change from hunter green, to apple green, to white specked with yellow, and so on throughout the visible spectrum, much like the South African chameleon Poulton described. But, as in the rotoscoping technique used to animate Philip K. Dick's scramble suit the mode of color change was the artist-technician's own painstaking application of filters, gels, and paint strokes. In the nonfilmic world, where there was no animator to maintain such "colour continuities and chromatics," to use Lye's words, even the best-made sniper suit could not match this performance. Only when its wearer leveraged the leaves and brush scattered in the environment and moved swiftly in the same direction as the wind could the suit help effect concealment in multiple environments.

Skin as Screen

Echoes of the challenges of representing chameleonic transformation in the early twentieth century are found in the attempts to effect a comparable transformation of the filmic medium and of the human skin. Not quite comfortable in their own skins, humans strive to supplement them with ideas, images, and technologies drawn from nature.

For Thayer, the primary conduit and medium of this grafting of natural protective concealment onto man was the photograph. This second skin was a photographic form, a cloth crafted from pieces of the photographic environment in which one was situated. Thayer proposed that man might wear such a cloth to disappear into a particular environment that was itself understood to be experienced primarily through visual, and specifically photographic, documentation.

Thayer's suiting method, however, worked only if the position of both observer and observed were understood to be fixed in place. If the observer moved a bit to the left, she would see through the static camouflage scheme. In a sense, then, Thayer provided a mode of concealment for flat people not unlike the "flat daddies," cardboard photographic figures that have recently appeared on the American home front as tokens of fatherly bodies in absentia.[26] A "flat daddy" is a stenciled silhouetted reminder to three-dimensional families of the "hole" in their home's visual, tactile, and emotional environments.[27]

But it is the men stationed overseas in hostile, dynamic environments who are the ones in need of a technology for human chameleonic concealment in the presence of an enemy's visual, infrared, and thermal surveillance devices. And yet effective chameleonic camouflage has remained out of reach. Just as our skin remains our own, Poulton's "rapid adjustable protective resemblance" remains a longstanding aspiration, rather than a practical reality of human military logistics.

But since the outset of the twentieth century, scientists, artists and writers have kept returning to the question of how animal powers of dynamic invisibility might be put toward human ends. How could humans seize—or at least create a technology to mimic—the control over real-time strategic concealment possessed by certain fish, lizards, and cephalopods? How might acquisition of a chamele-onic skin work in relation to a predatory surveillance that is photo-graphic in nature?

Most attempts to do precisely that start with study of the physi-ological mechanism of color change in these animals. As we've seen, Poulton speculated that the mechanism that controlled cha-meleonic color change involved direct connections between the animal's optic nerve and its epidermis.[28] These pigment cells, called chromatophores, became the site of fascination, especially as new color film stocks and processing technologies proliferated. The idea that humans would actively strive to harness or mimic these powers of chameleonic concealment was clearly expressed in turn-of-the-nineteenth-century literature. In H. G. Wells' science fiction novella *The Invisible Man* (1897), for example, invisibility—in contrast to its earlier utopian descriptions—was shockingly corporeal. In the novella, a policeman is startled during his first confrontation with the small town's peculiar interloper:

> "Why!" said Huxter, suddenly, "that's not a man at all. It's just empty clothes. Look! You can see down his collar and the linings of his clothes. I could put my arm—" He extended his hand; it seemed to meet something in mid-air, and he drew it back with a sharp exclamation. "I wish you'd keep your fingers out of my eye," said the aerial voice, in a tone of savage expostulation. "The fact is, I'm all here—head, hands, legs, and all the rest of it, but it happens I'm invisible."[29]

Mark Twain also brought up the human aspiration to imitate the capabilities of the chameleon in "The Chameleon in the Hotel Court" (1899), included in his *Following the Equator: A Journey Around the World*, published only a few years before he placed an order to acquire one of the hand-cranked "disappearing models" from Thay-er's workshop.[30] In *What is Man?* (1906), Twain even concluded that man "is a chameleon, by the law of his nature, he takes the color of his place of resort."[31]

Twain seems here to be making a mere metaphorical connection between the biological adaptation of a chameleon and man's social

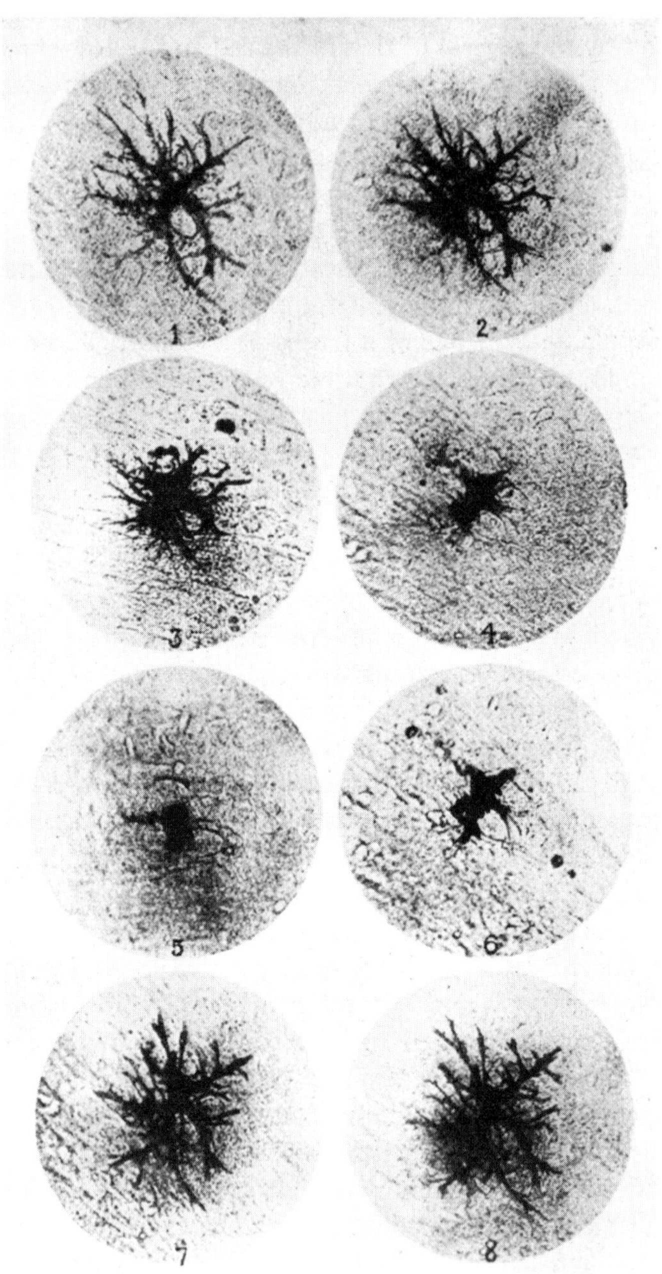

Figure 4.5—Microphotographs. Eight serial photographs made through a microscope documenting phases of expansion and contraction in a single chromatophore of a member of the Palaemonetes genus of freshwater shrimps. Reprinted from G. H. Parker, "Chromatophores," *Biological Reviews* 1 (January 1930), p. 68.

aspirations. Yet in the context of a general cultural attraction to biomorphism, the metaphor suggests a deeper desire as the real basis of the comparison: to collect, create, or wear an invisible skin—a chameleonic film—as one's body dances imperceptibly through a dynamic environment. Could such a skinly film be made of chromatophores (figure 4.5)?[32]

The netting of World War I was intended to be shape-shifting and adaptable—a crude prototype, perhaps, for more advanced technologies that transport camouflage logic into the human realm. The scout literally wrapped netting around his body. Veils, sniper suits, and netting were combined with natural materials "by using grass, leaves, etc. and by smearing hands and face and kit to harmonize with the surroundings."[33] But avoiding detection by the periscope on the ground and the aerial camera above was not a straightforward process. As Adrian Cornwell-Clyne, the officer in charge of experimental work for Army Camouflage, advised in a manual, "In order to attain perfect assimilation of the trench and men's heads with the surroundings, their color and texture must be identical with foreground and background. The silhouette of parapet and parados must follow a natural contour."[34] Out of sticks, cloth, and "camouflage consciousness" emerged an increasingly—but not perfectly—invisible man. This took hard work, constant attention, embodied projection, and self-analysis.

The familiar Disruptive Pattern Material, or DPM, that replaced the standard issue "Army green" infantry uniform since the Vietnam War was one attempt to arrive at a technology of effortless dynamic concealment that did the hard work for you.[35] After the United States Army, the Marines, Air Force, and Navy adopted it in the 1980s, other allied nations quickly followed suit with their own national DPM for use by soldiers of all ranks. "Garnish," however, like camouflage discipline, was still the responsibility of the individual.

The iconic DPM pattern, also known as Woodland M81, was developed at the laboratories of a prominent self-taught psychophysicist named Alvin Ramsley and patented in 1981 at the U.S.

Army Natick Soldier Systems Center in Natick, Massachusetts. It was an immediate hit. The pattern was printed on all sorts of personal accessories in the hopes it would help its wearers avoid both regular (human optical) and infrared detection systems operating from either ground or aerial perspectives.[36] Today, M81 is plastered on everything from hunting caps, to cigarette lighters, to bikinis.

Not on soldiers, however. In 2007, the U.S. Army replaced Woodland M81 with a digitized replacement, called the "universal camouflage pattern" or ACUPAT (Army Combat Uniform Pattern). The U.S. Marines switched from M81 to MARPAT (short for Marine Pattern), a patented digital camouflage pattern of interlocking multicolored "pixels" virtually identical to the army pattern. Both MARPAT and ACUPAT are patented and barred from civilian use. M81 has been deauthorized for American military use, but non-American armies the world over continue to use it in their standard-issue fatigues, sometimes even employing discarded or surplus American garments. Woodland DPM is, to some extent, an artifact of one scientist's engagement with the decisive need for making the individual tasks of camouflage easier. At the same time, it stamps members of an army with a badge of similarity. Its production—in a specific site-place-time-laboratory—was motivated by the aspirations of camouflage, that is, disappearance from photographic representation effected on the level of the individual or the battalion in the course of navigating a specific set of physical environments.

But it did virtually nothing to conceal the human form in arctic and desert terrains. Another reason for its decommissioning is that reconnaissance technologies were being developed for which M81 provided no countermeasure whatsoever. With the advent of stealth warfare, enemy reconnaissance via thermal detection was perceived as an increasing threat.

Predator, the 1987 action film directed by John McTiernan, stars Arnold Schwarzenegger as the head of an American Special Forces unit and vividly dramatized public concern about these emerging countersurveillance systems. It follows a team of snipers trained in techniques of personal concealment and enemy termination who

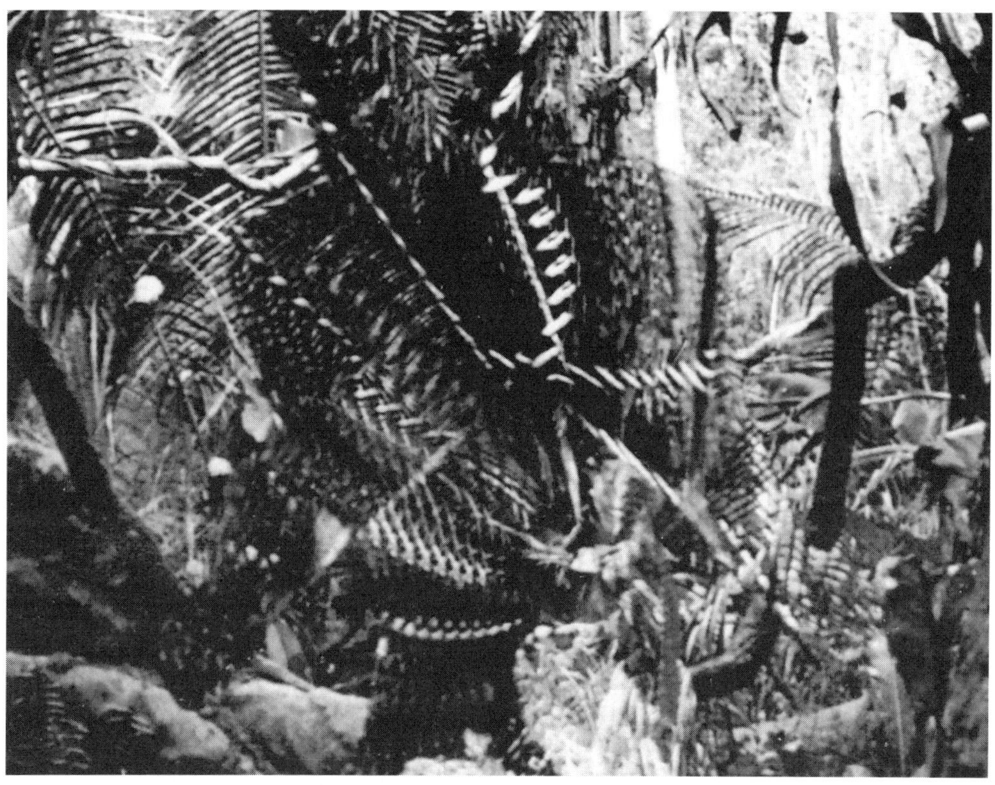

Figure 4.6—Predator effect. In this graphic based on a still from *Predator* (1987), the invisible hunter, played by Kevin Peter Hall, moves through the jungle and the film largely undetected. In this image from the film, Predator appears as a subtle outline in motion, moving through the scene as the jungle habitat bends around his form.

have been sent to the Latin American jungle ostensibly to recover an American kidnapped by guerilla rebels and Soviet military leaders. Once in the jungle, the Special Forces unit encounters another kind of enemy. They are unknowingly in the vicinity of an unseen creature, an "invisible man" who stalks and skins his prey. As he seeks out his human victims, the predator relies not on optics, but on a thermal detection system that translates heat signals into telltale shades of yellow, red, and blue. Schwarzenegger the hunter becomes Schwarzenegger the hunted.

Invisibility is the arbiter of tactical success within *Predator*'s narrative arc. At the outset, the Special Forces seem well equipped for jungle concealment. Schwarzenegger and his team are decked out in Woodland DPM uniforms and infrared night goggles. They have covered themselves in smudges of brown, black, and green that blur against the lush green and brown hues of the dense foliage.

The unseen creature, however, has both a mode of camouflage (an optical-electro chameleonic suit) and a means of surveillance (thermal detection) that the Special Forces lack and of which they are entirely ignorant until the end of the film (figure 4.6). Over the course of the middle section of the movie, viewers become aware of not only the predator's presence, but also his "filmic" style of hiding. His presence becomes visible in the form of "hiccups" in the texture of the film grain as it is pixilated along the contours of his silhouette. This occurs only at moments when he moves against the direction of the shot, as if the pixels are bent against the direction in which they want to move. When the predator is still and the camera is still, we see nothing. When one is still and the other is moving, we see nothing.

The predator's preternatural chameleonic camouflage, as pictured in figure 4.6, illustrates the ideal outcome of the optical-electro camouflage long anticipated as part of the U.S. Army's so-called Future Warrior System. The goal of optical-electro camouflage development is the production of a "wearable technology" for all terrains—fully adaptive strategic concealment. Optical and film signaling technologies are to be at the core of these technologies. In a perfected version of such a device, camera, screen, and projector would converge, fusing in a single sheet of photoreceptive fabric. This engineered fabric would be the basis for the chameleonic skin in which, like Lye's animated character in *Rainbow Dance*, a soldier could navigate variegated terrains, his veiling simultaneously recording a dynamic backdrop and projecting it into the photographic world. Camera, projector, and screen are rolled up into one skin. Together, they serve the purpose of chromatophores.

What the Department of Defense (DOD) currently aims to produce is identical in aspiration to the skin worn by the alien, Predator, in *Predator*.[37] Thus far, however, the DOD's prototypes are distinctly clunkier. As of 2003, a team of electrocommunications engineers reported on their innovation: "Optical camouflage uses the retro-reflective Projection Technology, a projection-based augmented-reality system composed of a projector with a small iris and retro reflective screen. The object that needs to be made transparent is painted or covered with retro reflective material. A projector then screens the background image on it making the masking object virtually transparent."[38]

Figure 4.7 depicts a diagram of one laboratory's optical-electrical camouflage system. The human subject stands in front of a video

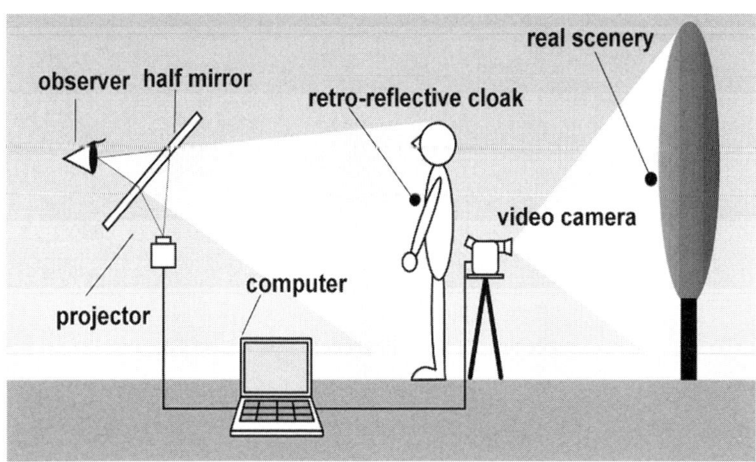

Figure 4.7—Retro-reflective projection technology. Reprinted courtesy of Tachi Laboratory, Keio University and the University of Tokyo.

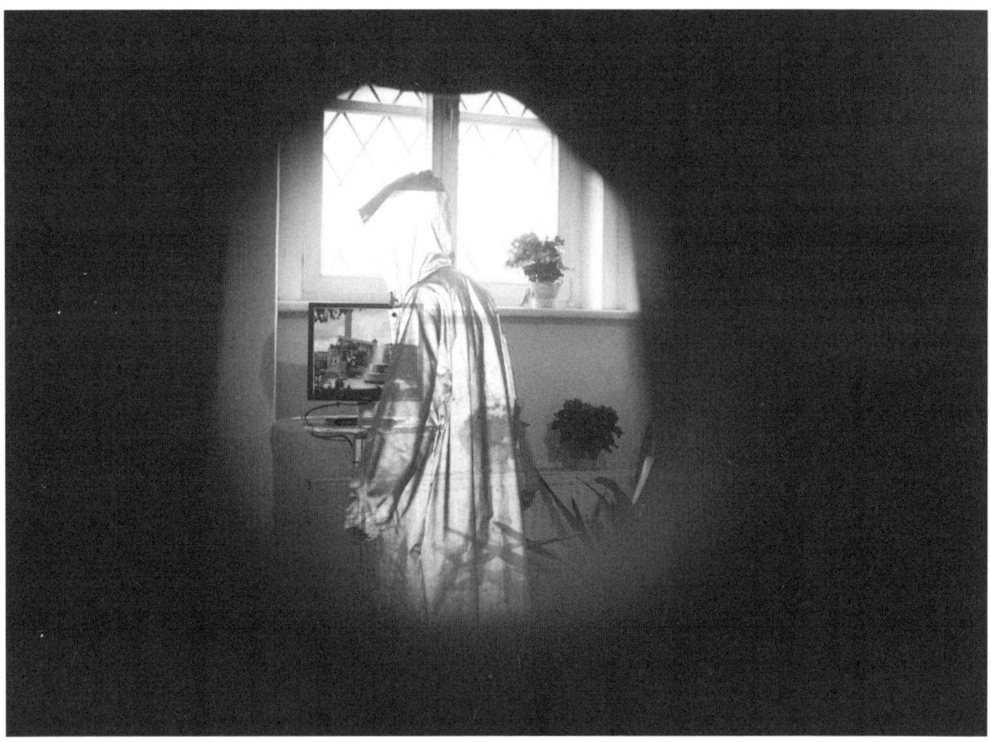

Figure 4.8—Adaptive optical camouflage. Photoreflective camouflage prototype. The man in this photograph stands before a video camera that collects a real-time feed of the image behind it and in turn links to a projector trained on a screen embedded in his tunic. Courtesy of Tachi Laboratory, Keio University and the University of Tokyo.

camera that collects a real-time feed of the image directly behind it. This camera is linked to a projector that casts infrared light directly onto the subject's tunic. This tunic—a textile skin operating as cinematic screen—is made of a photo-reflective material that converts infrared light to the visible spectrum, thereby reflecting light waves back to the observer. The photograph in figure 4.8 documents the effect thus produced. Although a remarkable prototype, the system's mechanism is complex, the technology cumbersome, and the concealment only partial. For example, the human subject must remain in close proximity to the camera, which has itself been optically conjugated with the observer's eye. As a result, far from being a potential secret agent navigating a dynamic environment, this "invisible man" is fixed in one place; he is a twenty-first century incarnation of the invisible men of Thayer's static photocollages. The photograph at left, then, simultaneously conceals and reveals the practical failure of the system it represents.

Camouflage media continuously adapt, whether in relation to static photography, serial photography, dynamic film/video systems, or emerging detection technologies. Advanced chameleonic media may someday allow the individual soldier or civilian to hide without having to anticipate how she might be detected—visually or otherwise. But so long as these technologies remains out of reach, or are promptly rendered obsolete by new forms of reconnaissance, we hide and seek simultaneously.

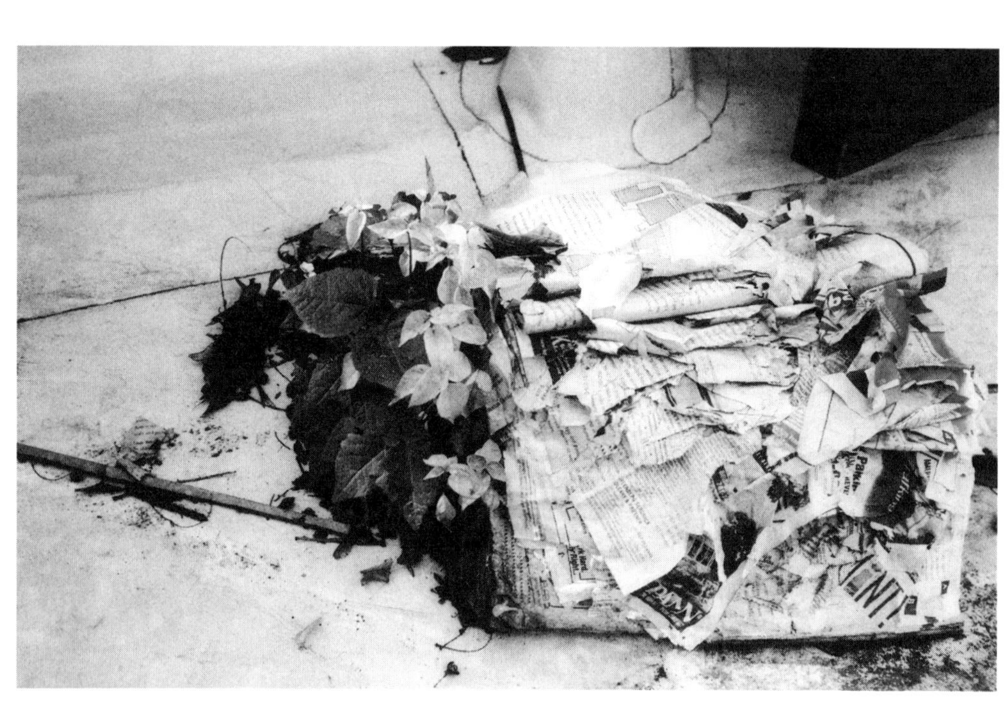

Acknowledgments

In the best of cases, words turn into other things. They allow other kinds of ideas to grow around them and on top of them: not just other peoples' words, but also new relationships, technologies, even forms of life (such as oak trees, as in this photograph, made of an object I found on the roof of my apartment building during the period I was writing the introduction to this book). And words themselves, in turn, grow from these other things—from sights, feels, smells, and the sounds of the words of others.

I could not have written this book without the creativity and wisdom of those around me. Thanks to my dissertation advisor, Peter Galison, and to Katharine Park, for fifteen years of mentorship, inspiration, and intellectual guidance as well as to dissertation committee members Everett Mendelssohn, Alexander Nemerov, Jennifer Roberts, and David Rodowick. Etienne Benson, Jimena Canales, Alex Csiszar, Michael Fischer, Stefan Helmreich, Martin Jay, Caroline Jones, David Jones, David Kaiser, Robin Kelsey, Vincent Lepinay, David Mindell, Heather Paxson, Natasha Schull, Sherry Turkle, Laura Wexler, Rosalind Williams, and Nasser Zakariya read full drafts of *Hide and Seek*, all providing insightful feedback. I am deeply grateful for the expertise, enthusiasm, and profound generosity of Roy Behrens.

My colleagues at MIT have provided an excellent environment for thinking, for writing, and for talking things through. Merrit Roe Smith and Clapperton Mavhunga taught me about guns. Harriet Ritvo brought back nonhuman animals and Leo Marx made me rethink everything. Chris Boebel, Mike Fischer, Stefan Helmreich,

William Urrichio, Chris Walley, William Urrichio, and Emily Zeamer supported this project's artistic scaffolding. Beyond MIT, Mark Allen, Zoe Beloff, Lisa Cartwright, Megan Dickerson, Luke Fischbeck, Alfred Guzzeti, Dan Hisel, Lana Z. Kaplan, Carolyn Lewenberg, Robb Moss, Sarah Rara, Jed Speare, and Carlin Wing have contributed to the the critical media practices that are central to both the argument expressed herein and to my larger mission as a scholar and artist. They have supported *Hide and Seek*'s manifestions as an experimental film, a performance piece at the Harvard Museum of Comparative Zoology, an interactive installation on Bumpkin Island in Boston Harbor, and an exhibition and workshop series at Machine Project in Los Angeles. Others whose various engagements with *Hide and Seek* have been invaluable include Ellen Bales, Eve Blau, Janet Browne, Matthew Christensen, Lorraine Daston, Bernard Geoghegan, Oliver Geycken, Orit Halpern, Amy Johnson, Joseph Koerner, Rebecca Lemov, Sharonna Pearl, Chitra Ramalingam, Meg Rotzel, Jennifer Tucker, Winnie Wong, and Josh Yumibe.

Many have contributed in particular and sometimes peculiar ways. Roger Hanlon taught me about cephalopod color change, and Peter Godfrey Smith introduced me to the philosophy of octopuses and made me want to learn to scuba dive. At the Harvard Society of Fellows, Dan Aaron gave me nearly a century of wisdom, Elaine Scarry helped me see colors anew, Bertrand Halperin taught me about invisible cloaking, and Wally Gilbert about the evolutionary genetics behind industrial melanism. Daniel Kronauer and Naomi Peirce gave me a tutorial on insect mimicry, allowing me to linger amid the Lepidoptera. Colin Kennedy, Tamara Sussman, and the intrepid participants in my art projects, made and donned ghillie suits. I thank Sina Nafjali, who published a version of the first chapter in the journal *Cabinet* in 2009. For her way with words, I am grateful to Megan Hustad, who has been both an aid and an inspiration. Jeremy Blatter and Melissa Lo assisted with research and Megan Houser, Gina Badger, and Chris Cammet assisted with graphics. Sarah Lawrence College, the Film Studies Program at the University of Chicago, the Center for New

Media at Berkeley, the Department of War Studies at King's College, the History of Science Society, and the Society for Cinema and Media Studies have provided amazing environments in which to present this work as it has evolved; I have gained much, and learned from the communities of curious minds in attendance. For all kinds of things, thanks are due to Odilia Bonebakker, Sarah Bruner, Lois Choi-Kain, Sonny Cytrynbaum, Danielle Mancini, Andrea Meld, Marc, Susan, and Jacob Shell, Judy Spitzer, Annemarie Strassel, and especially Jason Sanford.

Historical materials from many archives and museums are key to the spine and integuments of this book. Curators of texts, films, photographs, uniforms, weapons, and natural history specimens opened their doors, boxes, and file cabinets for *Hide and Seek*. I am grateful to curators at the Imperial War Museum, the Public Records Office, the London Natural History Museum, the Royal Academy of Art, the United States National Archives, the Smithsonian American Art Museum, the Archives of American Art, and the Harvard Museum of Comparative Zoology University. Several archivists and curators went far beyond the call of duty, making my research a special pleasure. At the Imperial War Museum, I especially thank Matthew Lee of the Film and Video Department, Matthew Moody, formerly of the Art Department, Martin Boswell of the Uniforms and Munitions Collection, Alan Wakefield of the Photography Department, and Alan Richards of the Documents Department. Thanks also to Maggie Reilly at the Zoology Museum at the University of Glasgow and Stella Brecknel, librarian at the Hope Library at Oxford University. At the Smithsonian American Art Museum, Denise Wamaling and Leslie Green (Department of Graphic Arts) and James Concha (Department of Painting) were especially helpful. At the Archives of American Art, Barbara Aikens and Liza Kerwin made all the difference. I am grateful to Alison Pirie and Jeremiah Trimble of the Museum of Comparative Zoology at Harvard University. The excellent librarians at Harvard's Ernst Mayr Library at have been wonderful to work with over the years; I am deeply grateful to Ronnie Broadfoot and Mary Sears. Thanks to

Herb Jones at the Dublin Historical Society and Tyler Cann, curator of the Len Lye Archive at the Govett-Brewster Art Gallery in New Plymouth, New Zealand.

Generous support was provided by the Andrew W. Mellon Foundation, the Institute for Historical Research at the University of London, the Graduate School of Arts and Sciences at Harvard, the Graduate Society, the Center for Military History of the United States Army, the Minda de Gunzberg Center for European Studies, the William F. Milton Fund, the Charles Warren Center for Studies in American History, the Jacob K. Javits Foundation at the U.S. Department of Education, the Max Planck Institute for the History of Science and the College Art Association's Millard Meiss Publication Fund. The Harvard Society of Fellows has graciously supported me in countless ways during the revisions of my manuscript. Thanks to everyone at Zone Books—editor Ramona Naddaff, managing editor Meighan Gale, general manager Gus Kiley, designer Julie Fry, and the inimitable copyeditor Bud Bynack.

For the guidance of my mentors, advisors, artistic collaborators, friends, and family, and for the insightful feedback of those who generously read, reread, and thoughtfully engaged with the manuscript, my gratitude is without bounds. *Hide and Seek* could not have appeared otherwise.

The Art of Disappearance

These plates document a journey through the development of the thought process of Abbott Thayer (1849–1921). They exhibit his aesthetic evolution over time and across media, between approximately 1890 and 1915. Implicit in Thayer's way of making is a unique understanding of how we acquire knowledge, true or false as that knowledge may be. Thayer engaged with the skins of animals, killing, preserving, and mounting them. He posed taxidermied birds in outdoor scenes and then presented them as hidden by necessity. The only way to ascertain the presence of an animal was to imagine its outline traced on the background, like the ducks cut out of wallpaper, who could be fit into their wallpaper background and therefore rendered invisible. Sometimes rather than mounting skins, he plucked their feathers. From these he assembled complex collages. Sent far and wide, exhibited at galleries and now all lost, these also became the inspiration for the series of paintings shown in plates 3 and 4. Stencils lain over paintings, and photo collages of apparently foregroundless natural scenes revealed figures that matched realist paintings of the same specimens. Humans, unlike all other animals, were not naturally hidden. By the first decade of the twentieth century, Thayer began to remedy this situation, interpreting tattoos and creating textile patterns to this end. Thayer's world was far different from the world most of us understand to exist. But how do we know, after all, that the world isn't full of hidden things? What predators, real or imagined, might be revealed by a stenciled outline?

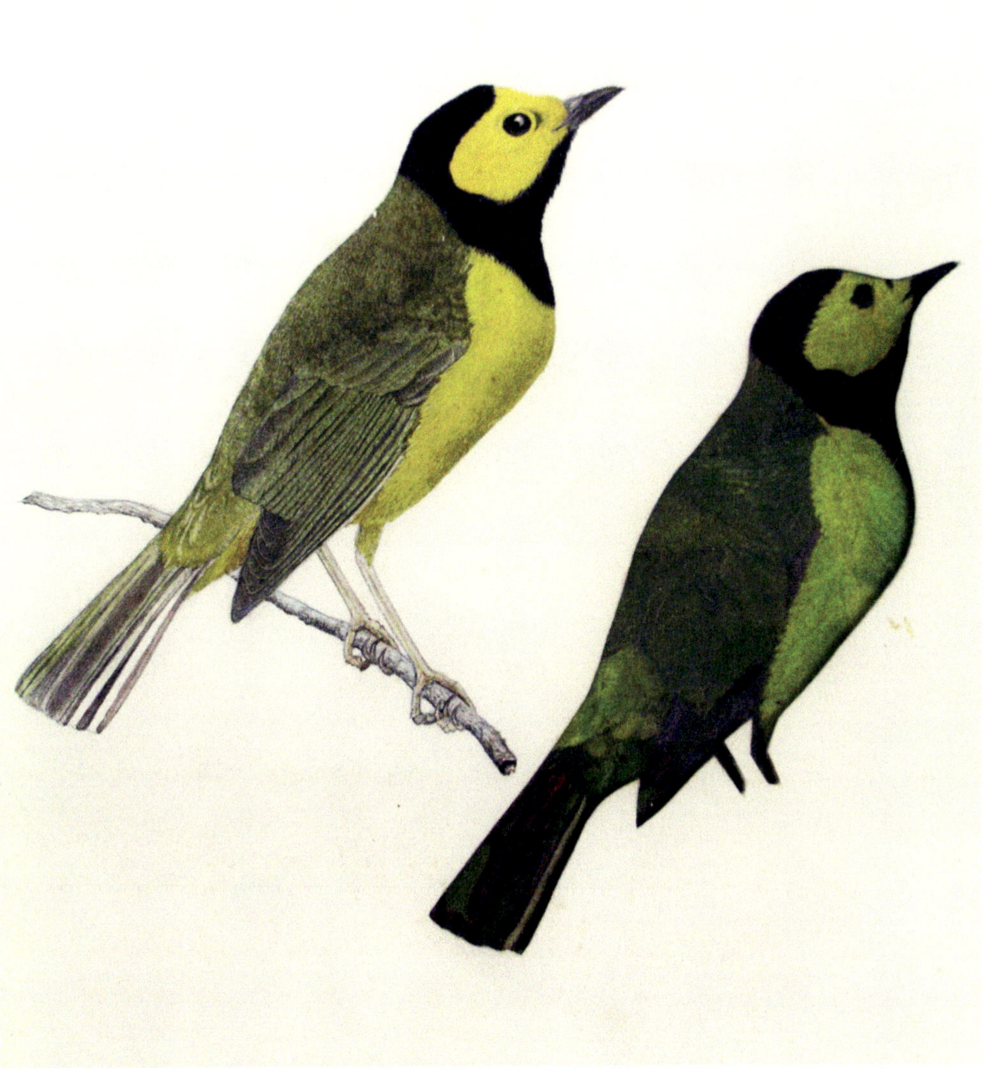

Plate 1—Hooded warblers: foreground. "The stenciled cutout in the shape of a real bird is used to produce a 'sham' creature of extraordinary likeness to the real one, when applied to the creature's background." See plate 16.

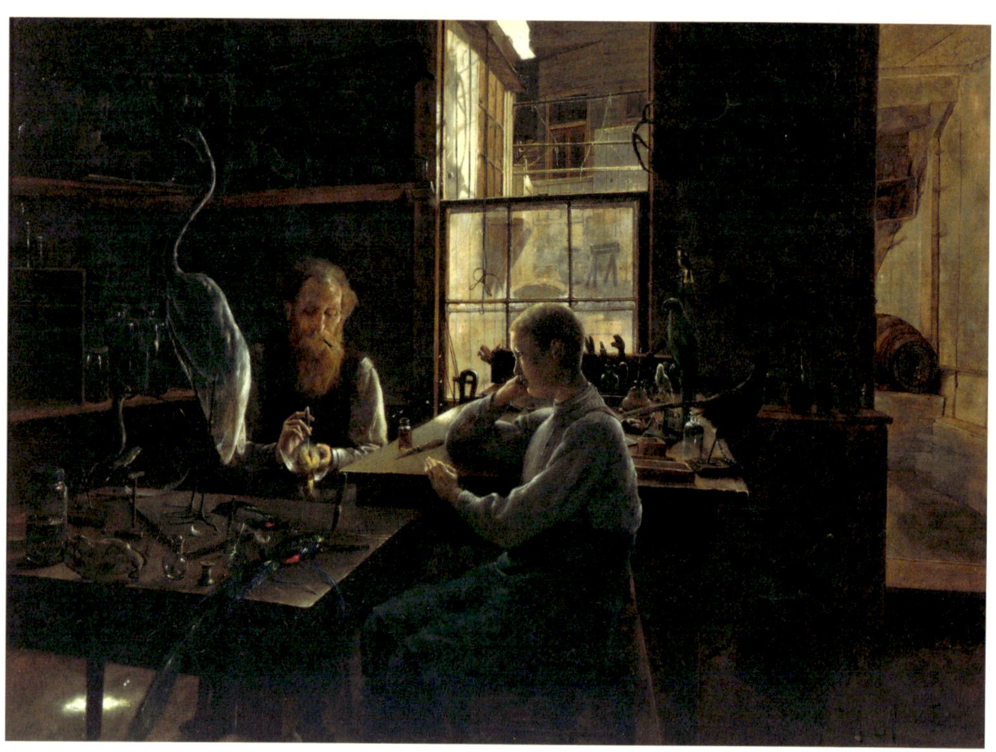

Plate 2—Young taxidermist. "He instructed me in the delicate art of enlarging a bird's rectum sufficiently to roll its skin tenderly back from its carcass. I then cut the carcass away from the skull and wings, put on rubber gloves, and rubbed the hide with powdered arsenic, rolled the skin back in place and stuffed the body with cotton."

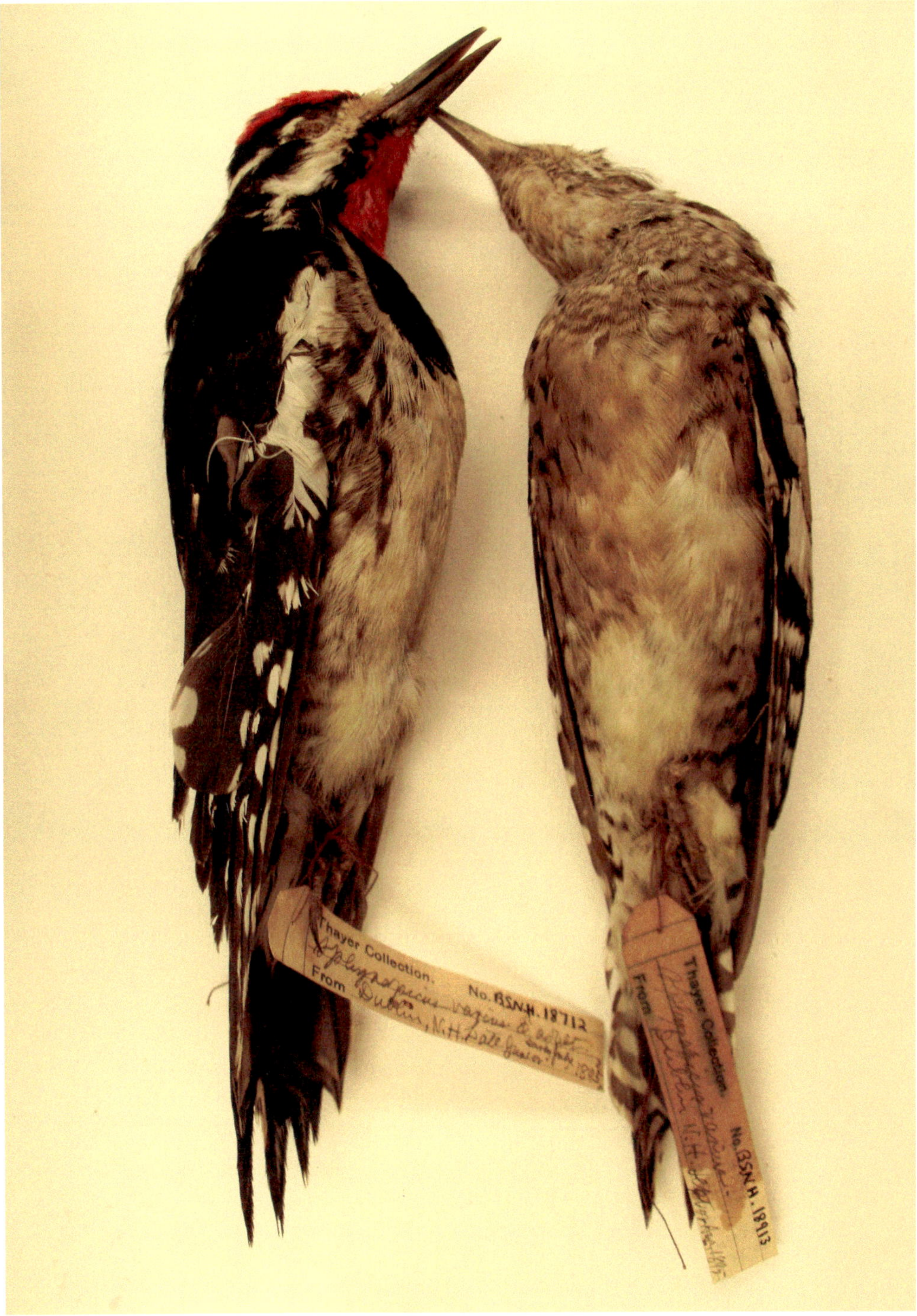

Plate 3—Thayer's specimens. Male and female yellow-bellied sapsucker (*Sphyrapicus varius*) collected, skinned, and preserved by Abbott Thayer in July and August 1895.

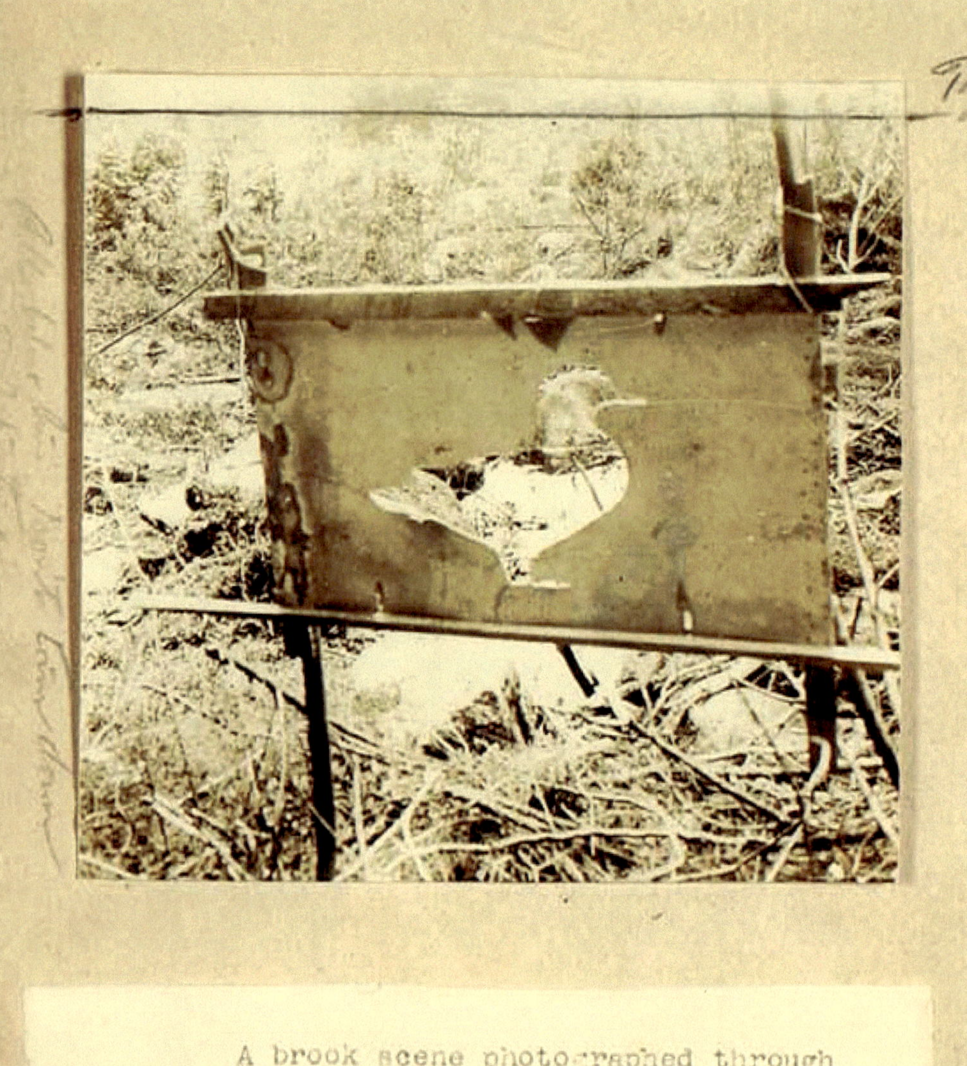

Plate 4—Photographic document. "I have alighted on the means of still more complete ocular demonstration." Photographing through a stencil reveals the (potential) presence of a duck that otherwise would be concealed from the camera's point of view.

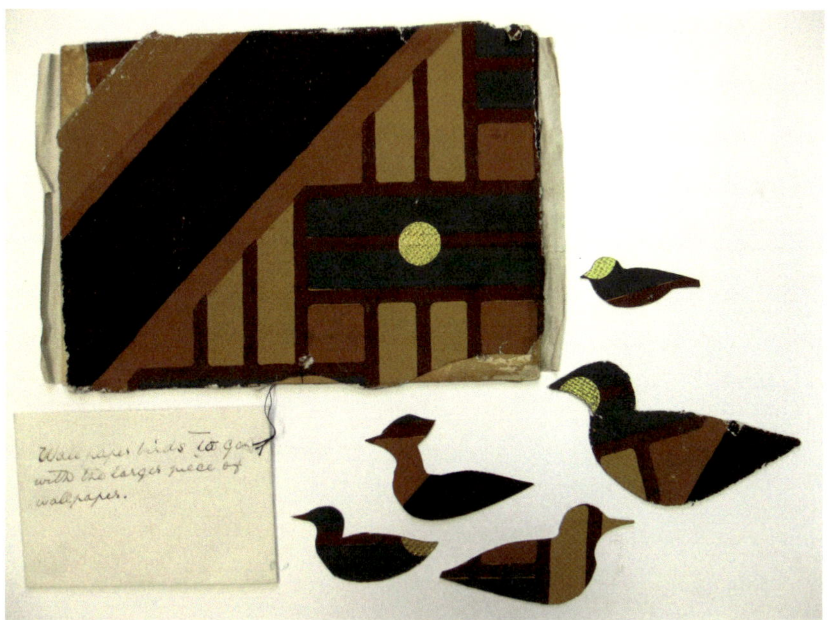

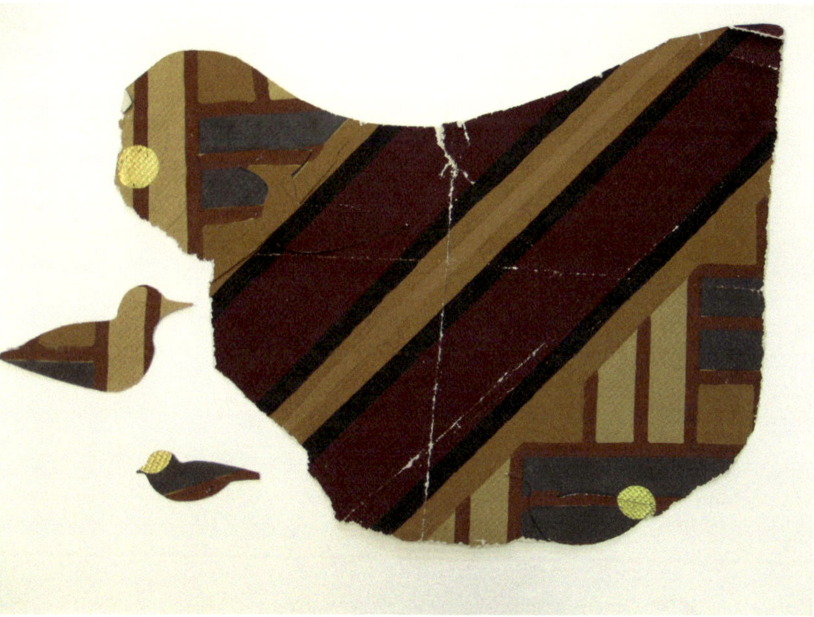

Plate 5—Wallpaper ducks. "I include extra wallpaper for more experimenting."

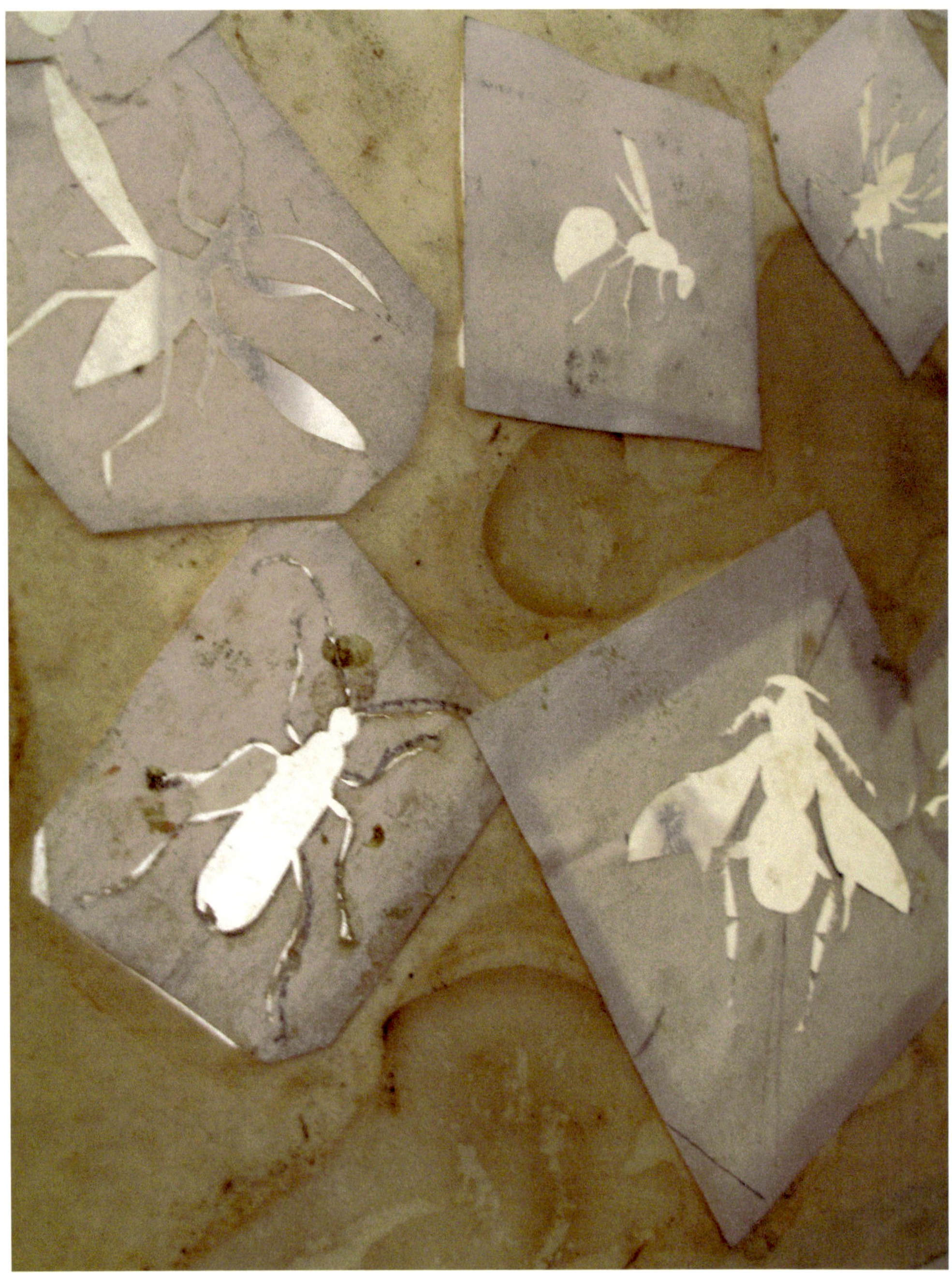

Plate 6—Wasps: finding aid. Placing this "finding aid" over the hand-toned photo collage (plate 7) reveals a series of wasps that perfectly match color illustrations from an insect identification handbook.

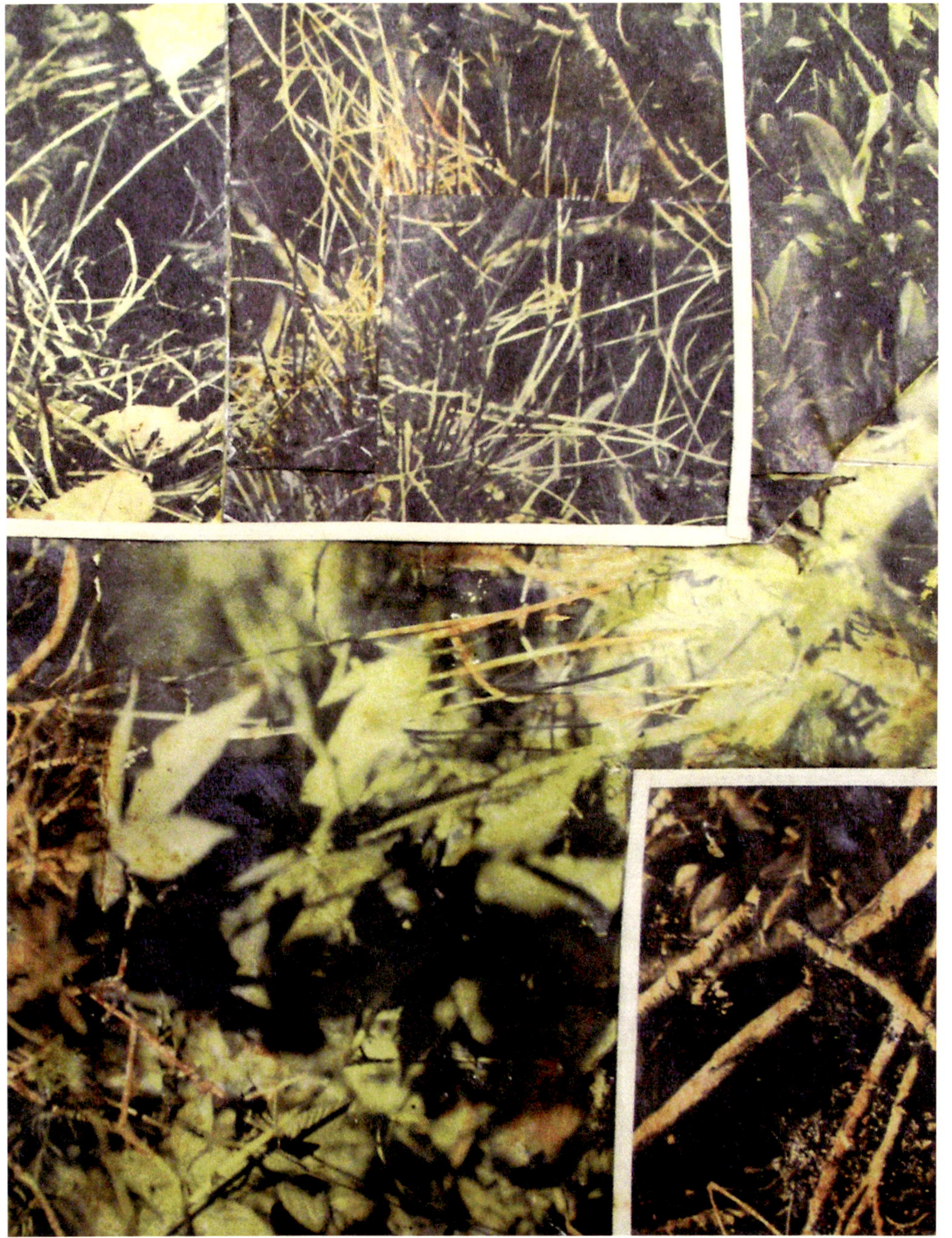

Plate 7—Wasps: photographic habitat. "Nature has to make, as it were, composite photographs of these animals' backgrounds true only to their average, but in this way they become truth itself—actual art!"

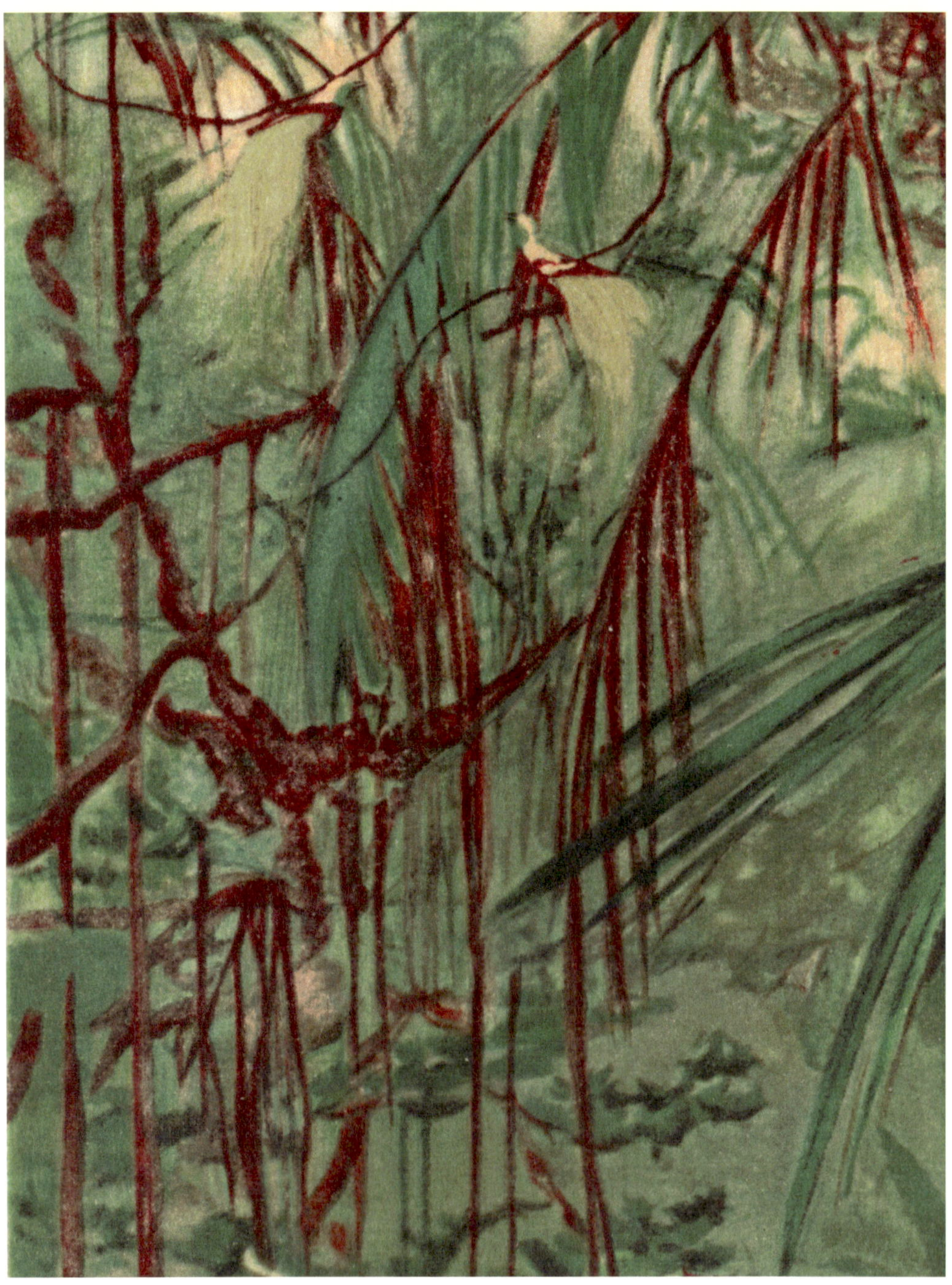

Plate 8—Feather painting. "I hope to send you in a few days the bird of paradise sketch made wholly out of this bird's three colors, to show absolutely how he, like probably all conspicuous birds, is an actual picture of his background."

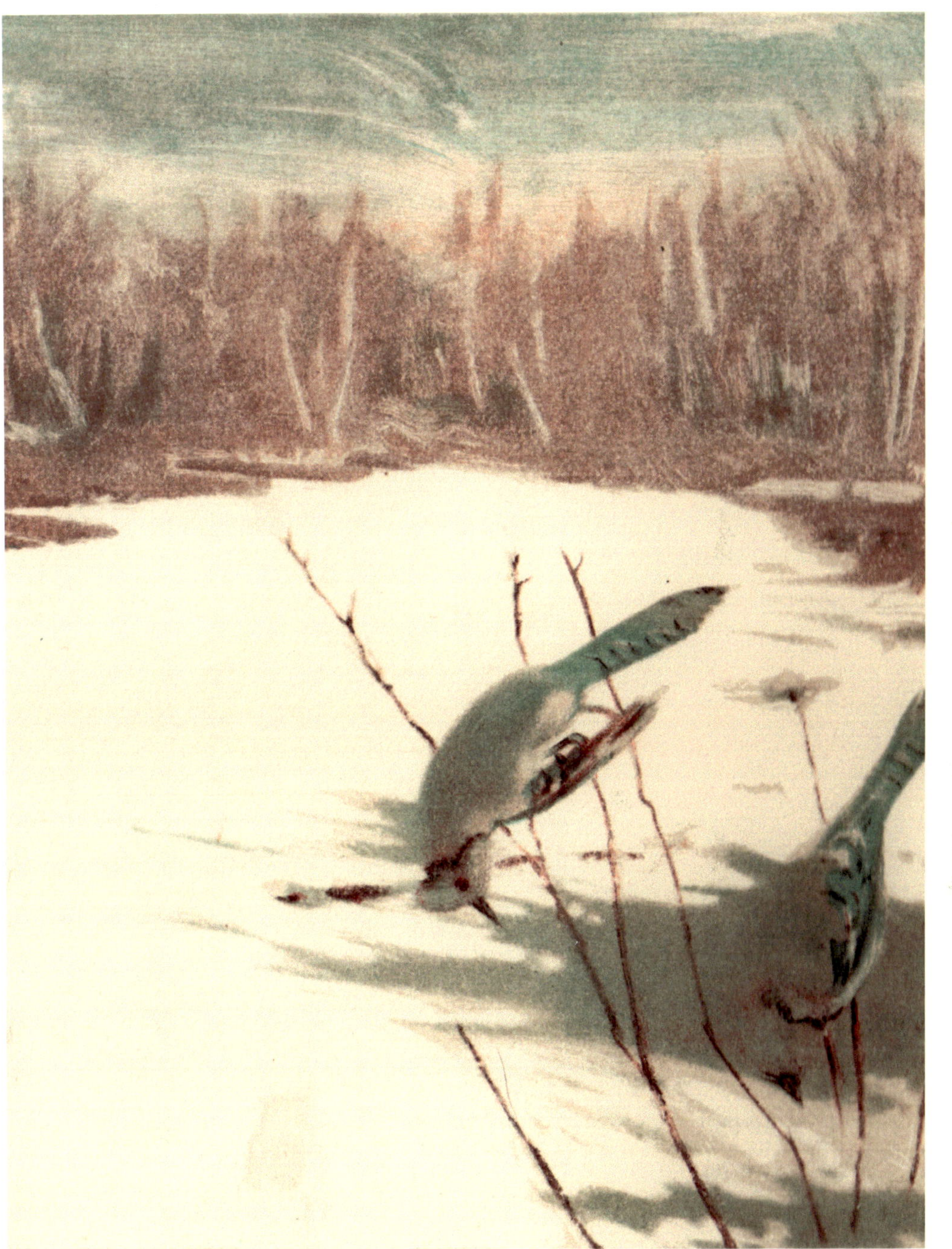

Plate 9— Painting stuffed birds. "My blue jay skin picture shows him to be a picture of the leafless season here."

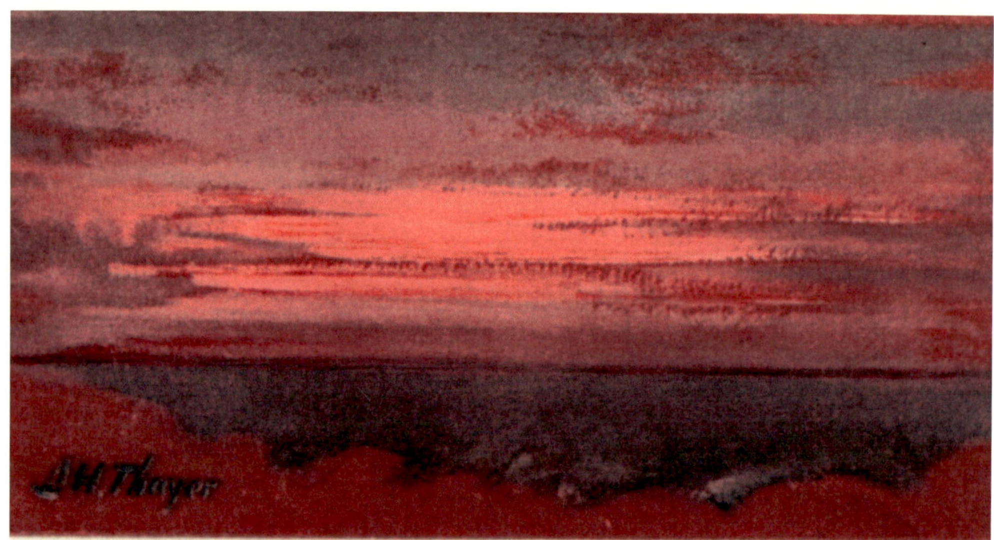

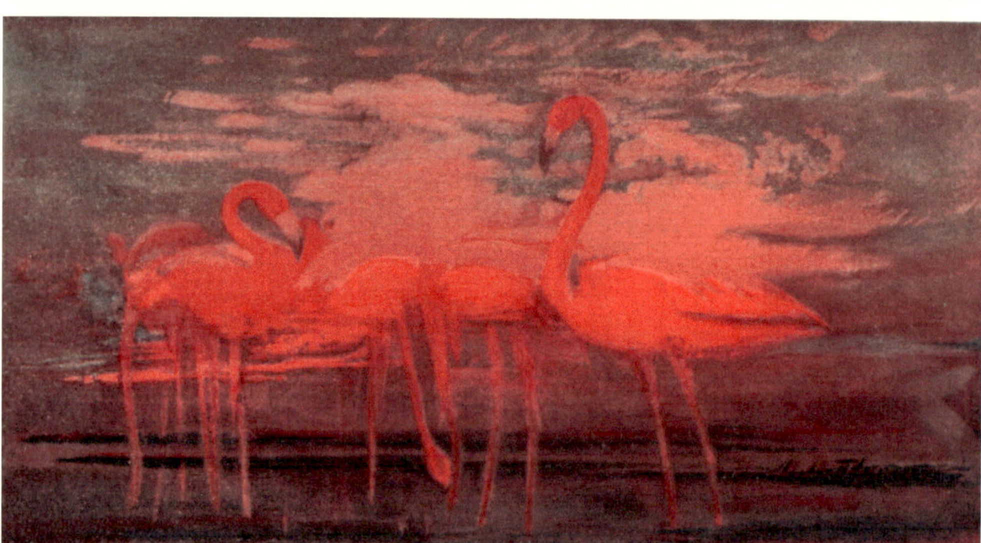

Plate 10—Sunrise or sunset. "How wonderfully such birds match or reproduce the colors of morning or evening skies to the eyes of the inhabitants of the water in which they wade."

Here is a three toned design. Cut a bird out of a part that has
only <u>two</u> of these three tones, and when you put him on a flat expanse
of the <u>third</u> tone, he <u>shows</u>. Cut a bird out of a part that gives him
a fair amount of _{each of} the <u>three</u> <u>tones</u> that compose the design, and even
when you put him on a flat expanse of <u>either</u> of the three tones (as
on the <u>red</u> expanse) he merely carries to that place the same illusion.

This principle is equally perfect in the effect of painting a <u>regi-</u>
<u>ment</u>. Find a place that contains plenty of all the main notes of the
region: <u>draw</u> <u>up</u> <u>your</u> <u>regiment</u> <u>there</u>, and efface them as completely as
this little pink, white and green bird is effaced!

Plate 11—How to paint a regiment. "Draw up your regiment there, and efface them as completely as
this little pink, white, and green bird is effaced."

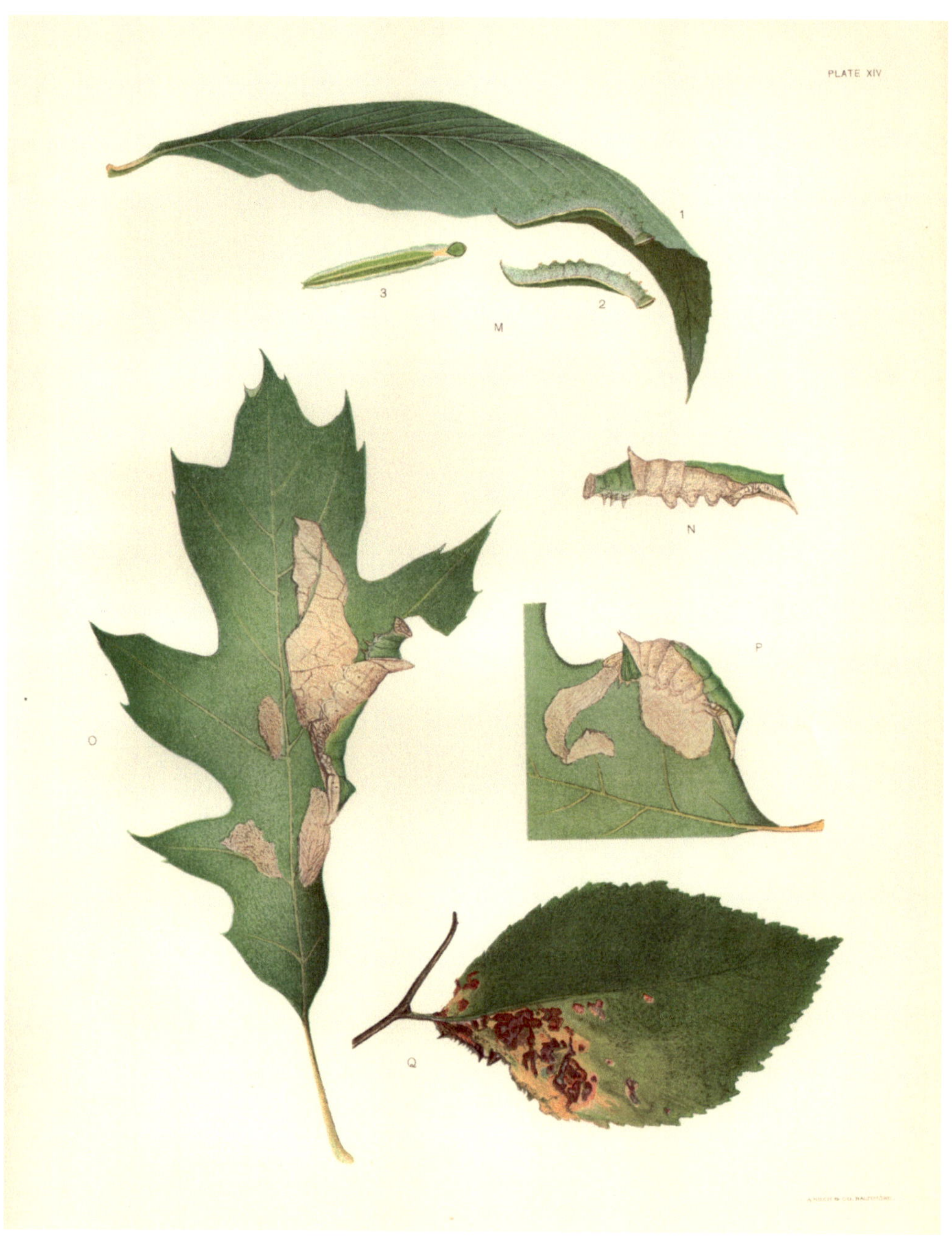

Plate 12—Leaves that deceive. Who can be sure that any foliage whatsoever does not itself consist of "decoy leaves"?

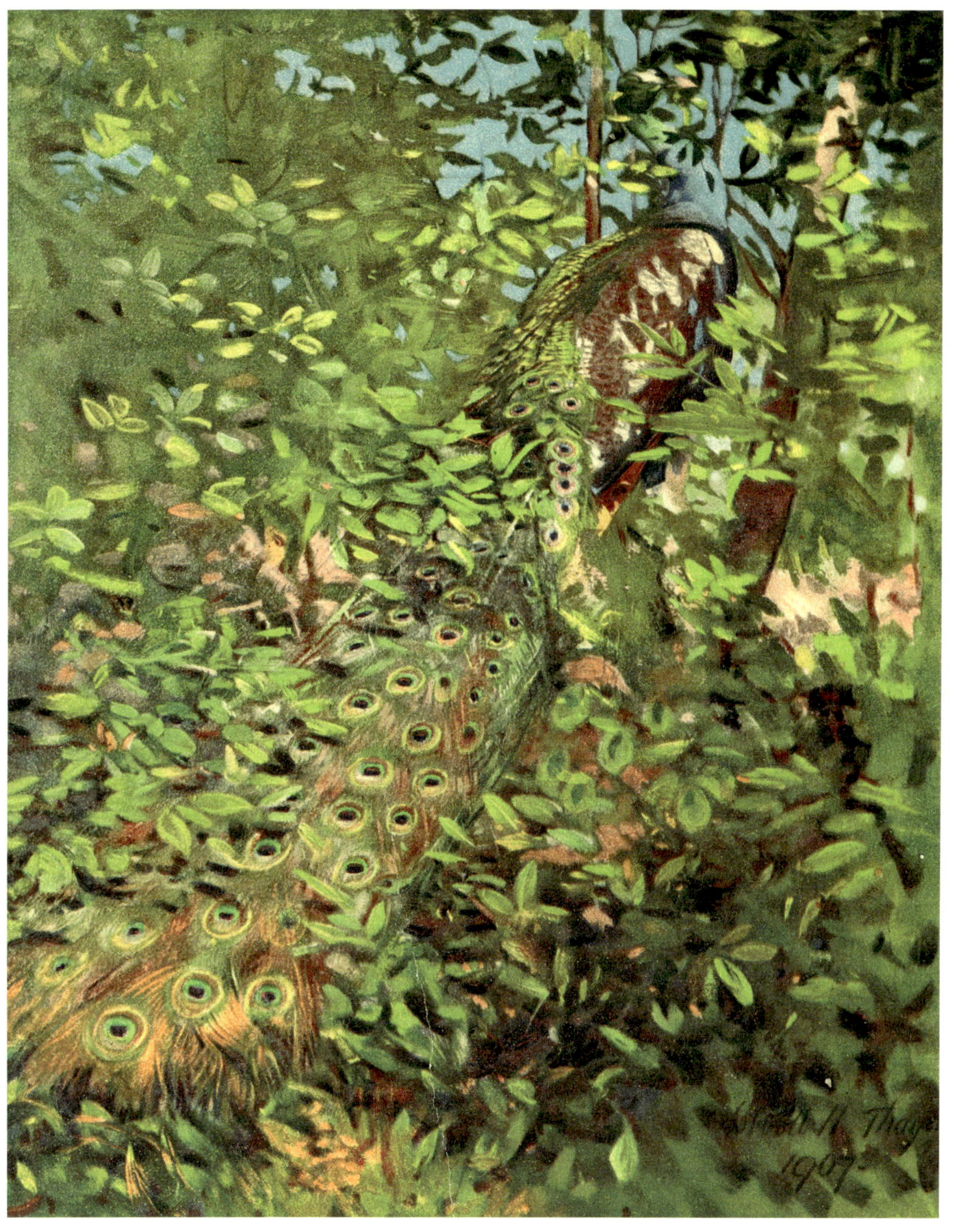

Plate 13—Peacock in woods. "The spread tail also looks very much like a shrub." See figure 1.14 and pp. 53–56.

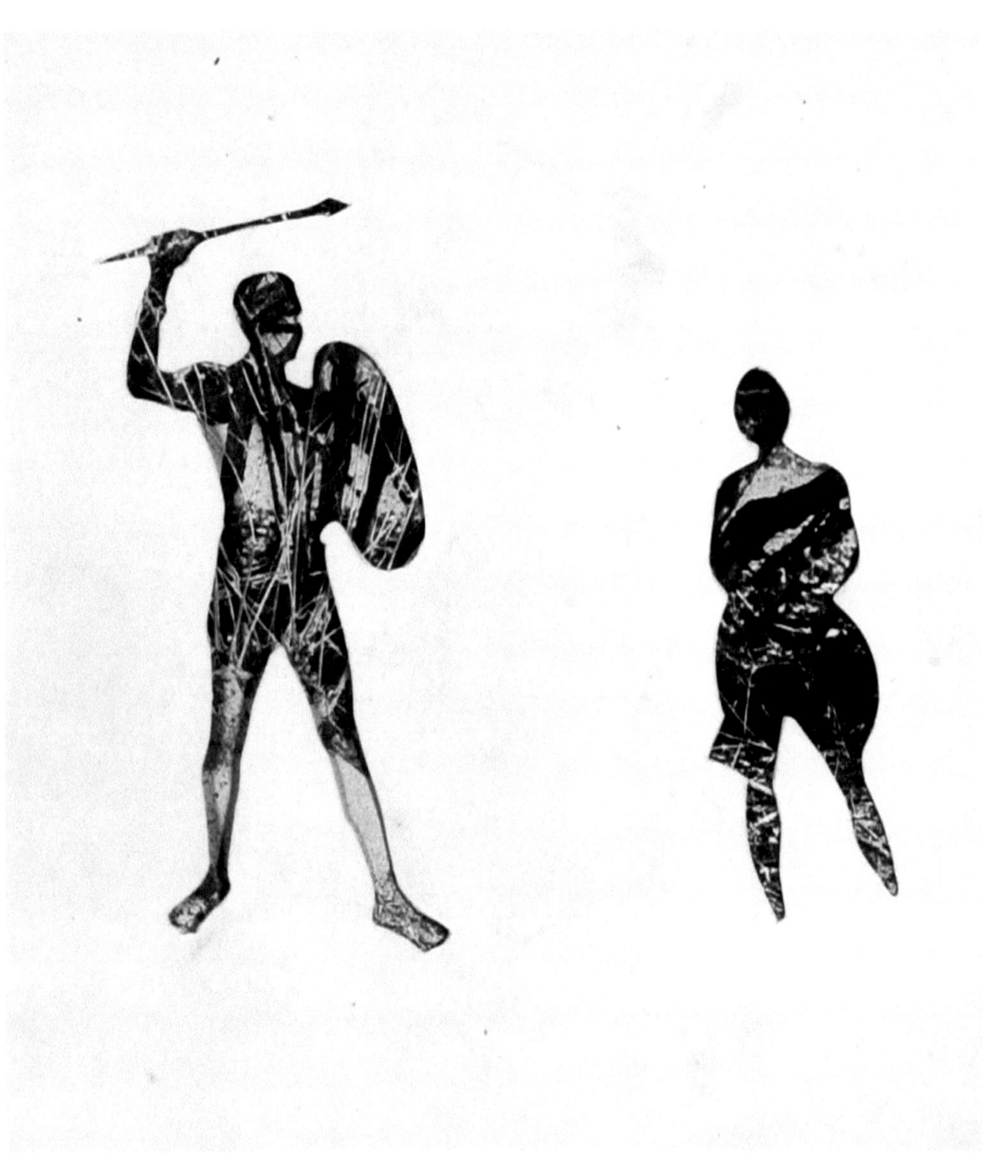

Plate 14—Two warriors. "Scene photographed through the stencil of a Kalinga warrior."

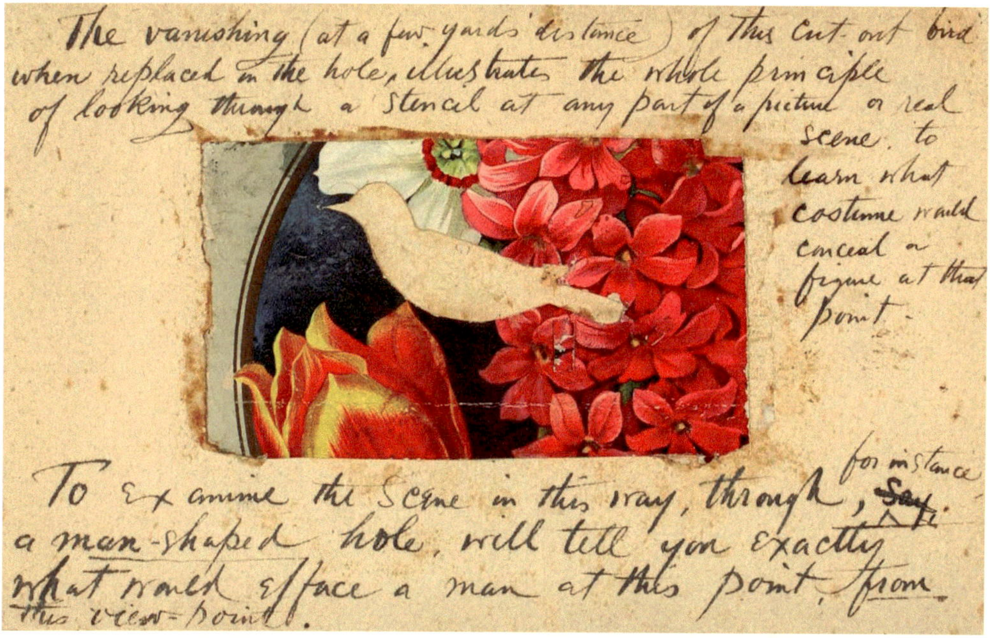

The vanishing (at a few yards' distance) of this cut-out bird when replaced in the hole, illustrates the whole principle of looking through a stencil at any part of a picture or real scene, to learn what costume would conceal a figure at that point.

To examine the scene in this way, through, for instance say a man-shaped hole, will tell you exactly what would efface a man at this point, from this view-point.

Plate 15—Man-shaped hole. "Man has only to cut out a stencil of the soldier, ship, cannon or whatever figure he wishes to conceal, and look through this stencil from the viewpoint under consideration."

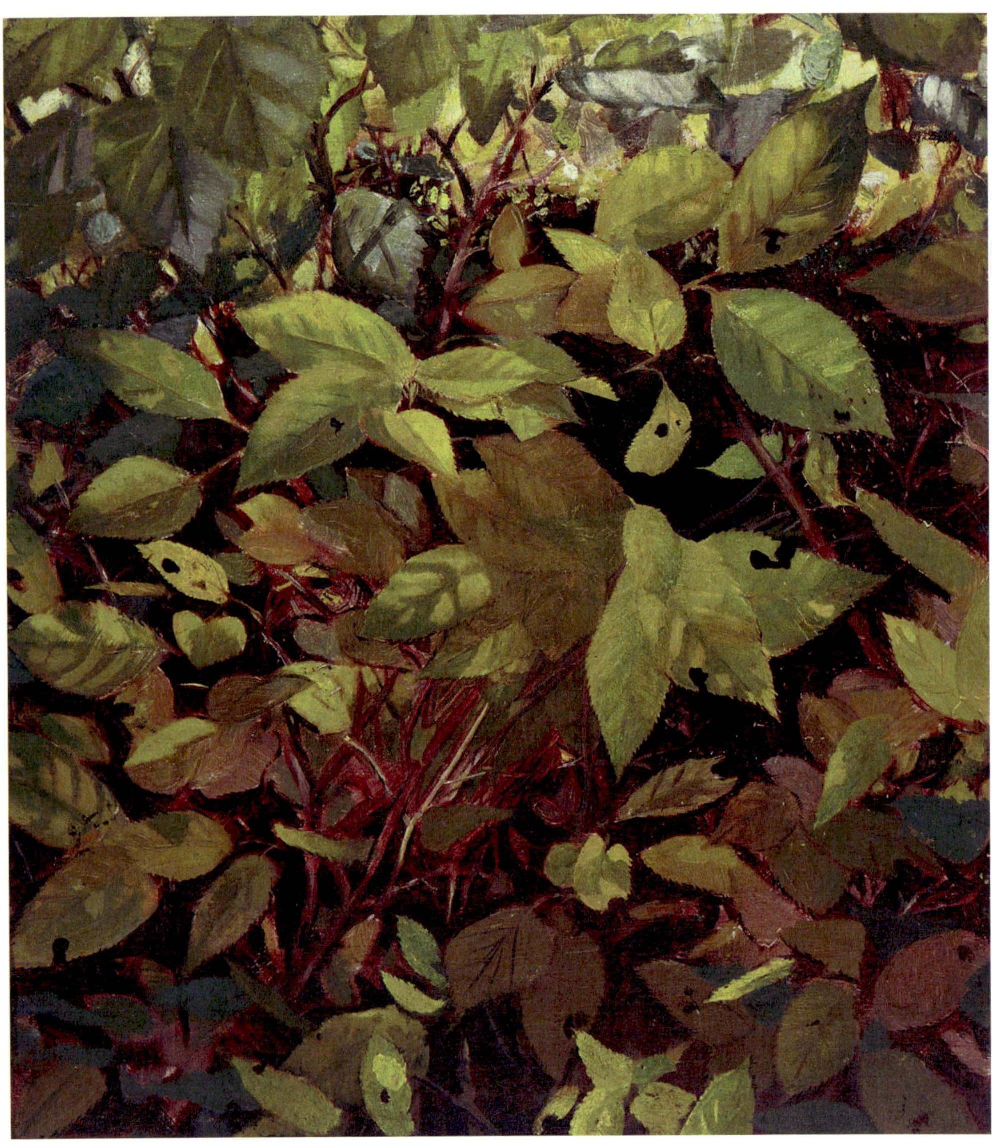

Plate 16—Hooded warbler: oil painting. "Oil painting, no feathers, hidden bird." See plate 1, figures 1.10 and 1.12, and pp. 50–53.

1. Reprinted courtesy of the Smithsonian American Art Museum [SAAM 1950.2.41]. Quotation from Abbott H. Thayer and Gerald H. Thayer, *Exhibition of Paintings Illustrating Protective Coloration in Nature* (New York: Pratt Institute Art Gallery, 1924), pp. 1–2.

2. Henry Alexander, *The First Lesson (The Taxidermist),* 1885. Reprinted courtesy of the Fine Arts Museums of San Francisco, Museum purchase, Mildred Anna Williams Collection [1952.76]. Quotation from Barry Faulkner, *Sketches from An Artist's Life* (Dublin, NH: William Bauhen, 1973), p. 43.

3. Courtesy of the Museum of Comparative Zoology, with thanks to Alison Pirie of the Ornithology Department. MCZ 29812 (left) and MCZ29808 (right). Photograph by the author.

4. Reprinted courtesy of the Smithsonian American Art Museum [SAAM 1950.2.36]. Quotation from Abbott H. Thayer, "Further Remarks on the Law Which Underlies Protective Coloration," *The Auk* 13 (October 1896), p. 318.

5. Reprinted courtesy of the Library of the Oxford University Museum of Natural History. Quotation from Abbott H. Thayer to Alfred R. Wallace, July 22, 1905, Library of the Oxford University Museum of Natural History.

6. Reprinted courtesy of the Smithsonian American Art Museum [SAAM 1950.2.54a].

7. Reprinted courtesy of the Smithsonian American Art Museum [SAAM 1950.2.54b]. Quotation from Abbott H. Thayer to Alfred R. Wallace, July 22, 1905.

8. Reprinted from *Concealing Coloration in the Animal Kingdom* (1909), p. 107. Quotation from Abbott H. Thayer to Alfred R. Wallace, July 22, 1905.

9. Reprinted from *Concealing Coloration in the Animal Kingdom* (1909), p. 107. Quotation from Abbott H. Thayer to Alfred R. Wallace, July 22, 1905.

10. "Sunrise or Sunset" and "Red Flamingos" reprinted from *Concealing Coloration in the Animal Kingdom* (1909), p. 156. Quotation from *Concealing Coloration,* p. 156.

11. Courtesy of the Smithsonian American Art Museum [SAAM 1950.2.32].

12. Caterpillars in watercolor, including "crumpled-and-withered-leaf-edge-mimicking caterpillar" and "oak-leaf-mimicking caterpillar," by Gerald H. Thayer and Gladys Thayer. Reprinted from *Concealing Coloration in the Animal Kingdom* (1909), p. 192.

13. Reprinted from the frontispiece to *Concealing Coloration*, based on "Peacock in Woods," an oil painting by Abbott Thayer assisted by Rickard Meryman. Quotation from original caption.

14. Courtesy of the Smithsonian American Art Museum [SAAM 1950.2.41]. Quotation from Abbott H. Thayer, "Camouflage," *The Scientific Monthly* 7.6 (Dec. 1918), p. 488.

15. Courtesy of the Smithsonian American Art Museum [SAAM 1950.2.39]. Quotation from Thayer, "Camouflage," p. 494.

16. *Hooded Warbler and Sunlit Foliage* painted by Gerald and Gladys Thayer. Reprinted courtesy of the Smithsonian American Art Museum [SAAM 1950.2.41]. Quotation from Abbott H. Thayer and Gerald H. Thayer, *Exhibition of Paintings Illustrating Protective Coloration in Nature* (New York: Pratt Institute Art Gallery, 1924), p. 2.

Notes

PREFACE

1. I have shown this photograph, blown up to be wall-sized, as well as printed at the size reproduced in figure 1, to many lecture audiences. (British soldiers in 1916 and students at the Camouflage School would similarly have been presented with this image as a glass lantern slide projection, as well as in smaller-scale pamphlet form.) Invariably, among the members of each audience (some even including ex-Vietnam and World War II photoreconnaissance specialists) I've encountered a wide range of responses—suggestions and frustrations among them.

2. Lorraine Daston and Peter Galison, "The Image of Objectivity," *Representations* 40 (Fall 1992), pp. 81–128; Ian Hacking, *Representing and Intervening: Introductory Topics in the Philosophy of Natural Science* (Cambridge: Cambridge University Press, 1983); Joel Snyder and Neil Walsh Allen, "Photography, Vision and Representation," *Critical Inquiry* 2.1 (Autumn 1975), pp. 143–69; James Elkins (ed.), *Photography Theory* (New York: Routledge, 2007), pp. 129–205.

3. Lorraine Daston and Peter Galison, *Objectivity* (New York: Zone Books, 2007); Alan Sekula, "The Body and the Archive," *October* 39 (Winter 1986), pp. 3–64; John Tagg, "The Photograph as Evidence in Law," in *The Burden of Representation: Essays on Photographies and Histories* (Minneapolis: University of Minnesota Press, 1993), pp. 66–103; Lisa Cartwright, "Scientific Looking, Looking at Science," in Marita Sturkin and Lisa Cartwright (eds.), *Practices of Looking: An Introduction to Visual Culture* (New York: Oxford University Press, 2001), pp. 279–313; Jennifer Tucker, "Photographic Evidence and Mass Culture," in *Nature Exposed: Photography as Eyewitness in Victorian Science* (Baltimore: Johns Hopkins University Press, 2005), pp. 194–234.

INTRODUCTION

1. Barry Faulkner, *Sketches from An Artist's Life* (Dublin, NH: William Bauhen, 1973), p. 91.

2. "Seeing but Not Seen," *Scientific American*, May 18, 1918, p. 451. Also War Department, Camouflage School. CS-81. *Royal Air Force Camouflage Practice*, by Lt. Col. Homer Saint-Gaudens, August 13, 1942.

3. It was a "second nature" in two senses: in the first instance, as an acquired and internalized "habit" or consistent way of interacting with one's environment, and in the second instance, more literally, as simultaneously indexical and iconic, as above. Here I am drawing on William Cronon's distinction based on his reading of Marx, laid out in Cronon's introduction to *Nature's Metropolis: Chicago and the Great West* (New York: W. W. Norton, 1992), pp. 41–64.

4. United States Department of Defense, *Dictionary of Military Terms* (London: Greenhill Press, 1999), s.v "Camouflage." Paul Imbs, *Trésor de la langue française 1789–1960*, vol. 5 (Paris: Éditions Scientifique, 1977), p. 78. On *camouflet*, see Amédée Beaujean, *Dictionnaire de la langue française; abrégé du dictionnaire de É. Littré de L'Académie française avec un supplément d'histoire et de géographie, par A. Beaujean*, 8th ed. (Paris, Librairie Hachette, 1886) p. 159.

5. This is not unlike, in fact, the "operation post (O.P.) trees" developed by the British and installed along the front in World War I. (See Chapter 2.)

6. Homer Saint-Gaudens, "Camouflage and Art," *The Art Bulletin* 2.1 (September 1919), pp. 23–30.

7. Popular studies of camouflage that have appeared in recent years and that discuss many of these linked histories and meanings include Guy Hartcup, *Camouflage: A History of Concealment and Deception in War* (New York: Scribners, 1980); Roy Behrens, *False Colors: Art, Design, and Modern Camouflage* (Dysart, IA: Bobolink Books, 2002) and *Camoupedia: A Compendium of Research in Art, Architecture, and Camouflage* (Dysart, IA: Bobolink Books, 2009); Tim Newark, *Camouflage* (London: Thames & Hudson, 2007); Hardy Blechman, *DPM* (London: Frances Lincoln, 2009).

8. M. E. N. Mejerus, C. F. A. Brunton, and Brunto J. Stalker, "A Bird's Eye View of the Peppered Moth," *Journal of Evolutionary Biology* 13.2 (2000), p. 159.

9. Ellen Barry, "Finland Sees a Familiar Pattern in Photos From the Georgia Conflict," *New York Times*, November 21, 2008, p. A6.

10. Paintings made in 1986; see Brenda Richardson, *Andy Warhol: Camouflage*

Paintings (New York: Gagosian, 1998).

11. Pamela M. Lee, "The World as Figure/Ground and its Disturbance," in Thomas Hirschhorn, *Utopia, Utopia = One World, One War, One Army, One Dress* (Boston/San Francisco, ICA/CCA, 2006), pp. 7–14.

12. Gertrude Stein, *Picasso* (London: B. T. Batsford, 1938), reprinted in Edward Burns (ed.), *Gertrude Stein on Picasso* (New York: Liverwright, 1970), p. 18.

13. Stephen Kern construes camouflage largely as a set of practices bound tightly with the development of cubism, on the one hand, and political modernism, on the other. Stephen Kern, *The Culture of Time and Space 1880–1918: With a New Preface* (1983; Cambridge, MA: Harvard University Press, 2003), pp. xxv and 8. See also Stephen Kern, "Cubism, Camouflage, Silence, and Democracy: A Phenomenological Approach," in Roger Friedland and Deirdre Boden (eds.), *NowHere: Space, Time and Modernity* (Berkeley: University of California Press, 1994), pp. 163–81. Elizabeth Louise Kahn, in *The Neglected Majority: "Les Camoufleurs," Art History, and World War I* (Lanham, MD: University Press of America, 1984), draws deeply on various biographical connections, emphasizing specific camouflage patterns while largely overlooking at aspects of camouflage so omnipresent as to literally become (almost) invisible, even in many cases to the historian. See also, Patrick Wright, "Cubist Slugs," *London Review of Books* 27–12 (June 23, 2005), pp. 16–20.

14. Roger Caillois, "Mimicry and Legendary Psychasthenia," (1935), trans. John Shepley, *October* 31 (Winter 1984), p. 31, a revised chapter of Roger Caillois's *Mythe et Homme* (Paris: Gallimard, 1938); Joyce Cheng, "Mask, Mimicry, Metamorphosis: Roger Caillois, Walter Benjamin and Surrealism in the 1930s," *Modernism/modernity* 16.1 (January 2009), pp. 61–86

15. Jacques Lacan, "Le stade du miroir: Théorie d'un moment structurant et génétique de la constitution de la réalité, conçu en relation avec l'expérience et la doctrine psychanalytique," *Communication au 14e Congrès psychanalytique international* (Marienbad: International Journal of Psychoanalysis, 1937); Jacques Lacan, "The Line and Light," in *The Four Fundamental Principles of Psychoanalysis*, ed. Jacques-Alain Miller, trans. Alan Sheridan (New York: W. W. Norton, 1981), pp. 91–104.

16. Paul Virilio, *War and Cinema: The Logistics of Perception* (London: Verso, 1989); Paul Virilio, *The Vision Machine* (Bloomington: Indiana University Press, 1994).

17. Virilio, *War and Cinema,* p. 18.

CHAPTER ONE: PRODUCTIVE MIMESIS
AND THE ART OF THE DISAPPEARANCE

1. "Fourteenth Congress of the American Ornithologists' Union," *The Auk* 14 (January 1897), 82–86.

2. Abbott H. Thayer, "The Law Which Underlies Protective Coloration," *The Auk* 13 (April 1896), pp. 124–29; Abbott Handerson Thayer, "Further Remarks on the Law Which Underlies Protective Coloration," *The Auk* 13 (October 1896), pp. 318–25.

3. Abbott H. Thayer, "An Essay on the Psychological and Other Basic Principles of the Subject," in Gerald Thayer and Abbott H. Thayer, *Concealing Coloration in the Animal Kingdom: An Exposition of the Laws of Disguise through Color and Pattern, Being a Summary of Abbott H. Thayer's Discoveries* (New York: Macmillan, 1909), p. 4.

4. "Thayer's Law" is still cited and debated in the fields of behavioral ecology and evolutionary biology. Examples include Graeme D. Ruxton, Michael P. Speed, and David J. Kelly, "What, If Anything, Is the Adaptive Function of Countershading?" *Animal Behaviour* 68.3 (September 2004), pp. 445–51; and Michael P. Speed, David J. Kelly, Andrew M. Davidson and Graeme D. Ruxton, "Countershading Enhances Crypsis with Some Bird Species But Not Others," *Behavioral Ecology* 16.2 (March–April 2005), pp. 327–34.

5. Thayer, "The Law Which Underlies Protective Coloration," p. 125.

6. Valuable accounts of Thayer's work, include Roy Behrens, "Revisiting Abbott Thayer: Non-scientific Reflections About Camouflage in Art, War and Zoology," *Philosophical Transactions of the Royal Society* 364.1516 (February 27, 2009), pp. 497–501. Nelson White, *Abbott H. Thayer: Painter and Naturalist* (Hartford: Connecticut Printers, 1951); Muriel Blaisdell, "Natural Theology and Nature's Disguises," *Journal of the History of Biology* 15.2 (Summer 1982) pp. 163–89; Sharon Kingsland, "Abbott Thayer and the Protective Coloration Debate," *Journal of the History of Biology* 11.2 (Fall 1978), pp. 223–44; Alexander Nemrov, "Vanishing Americans: Abbott, Theodore Roosevelt and the Attraction of Camouflage," *American Art* 11.2 (Summer 1997), pp. 50–81.

7. Frank Chapman, *Autobiography of a Bird Lover* (New York: D. Appleton-Century, 1933), pp. 78–79.

8. Reported by American Ornithologists' Union secretary J.H. Sage in "American Ornithologists' Union," *Science*, n. s. 4.102 (December 11, 1896), pp. 868–70.

9. Several of Thayer's students carried his methods and his worldview (which would only be granted the name "camouflage" in 1915, as we've seen in the previous chapter) into the thick of World War I. His students Homer Saint-Gaudens, Barry Faulkner, and George de Forest Brush, along with several others, founded the New York Camouflage Society in 1916, eventually the nucleus for the first American Camouflage Corps. "Wants Camouflage Force," *New York Times*, August 30, 1917, p. 13; "Faking as an Art in Conducting War: Fifty American Artists Form an Association for Camouflage to Aid the Army," *New York Times*, June 24, 1917, p. 35; H. Saint-Gaudens, "Camouflage and Art," *Art Bulletin* 2.1 (September 1919), pp. 23–30.

10. Thayer, "The Law Which Underlies Protective Coloration," p. 126.

11. Saroj Ghose, "From Hands-On to Minds-On," in Svante Lindquist (ed.), *Museums of Modern Science* (Canton, MA: Science History Publications and the Nobel Foundation, 2000), p. 118; see also, Bradford Washburn, "The Disappearing Exhibit," *New England Naturalist* (1939), pp. 20–21.

12. By 1909, Lucas had moved to the Brooklyn Museum and referred by letter to Thayer to a so-called "Disappearing Exhibit" by Thayer then under refurbishment and elaboration at the Brooklyn Museum. The presidents of both Wesleyan and Williams College requested models for their own university museums. For example, Frederic Lucas from the Smithsonian mentions models in his letters to Thayer of 1902. Abbott H. Thayer Papers, Archives of American Art (AAA).

13. W.B. Pyecraft, *Camouflage in Nature* (London: Hutchinson and Co., 1925); and *Abbott H. Thayer Exhibit Catalog*, Carnegie Institute, Pittsburgh, April 24–June 30, 1919.

14. Thayer translated his disappearing natural history exhibits into human terms. He installed a version of the duck exhibit based on miniature versions of the *Venus de Milo* and powered by electric lights in Dublin, New Hampshire. See illustration, ink on paper, in Nelson and Henry C. White Research Material, AAA. Also published in White, *Abbott H. Thayer*, p. 112.

15. Abbott H. Thayer, "Camouflage," *The Scientific Monthly* 7 (December 1918), pp. 481–94; A.M. Jungmann, "Dame Nature—Instructor in Camouflage," *Popular Science Monthly* 93.3 (September 1918), pp. 346–47.

16. George John Romanes, *Mental Evolution in Animals: With a Posthumous Essay on Instinct by Charles Darwin* (New York: D. Appleton, 1884), p. 324.

17. William James, *Principles of Psychology*, 2 vols. (1890; New York: Dover Publications, 1950), pp. 2:94–95.

18. Hanna Rose Shell, "Skin Deep: American Taxidermy, Embodiment, and Extinction," in Alan E. Leviton (ed.), *Museums and Other Institutions of Natural History, Past, Present, and Future* (San Francisco: California Academy of Sciences, 2004), pp. 88–112.

19. William T. Hornaday, *Taxidermy and Zoological Collecting* (New York: Scribners, 1891); Robert Shufeldt, *Scientific Taxidermy for Museums* (Washington, D.C.: Smithsonian, 1894).

20. William T. Hornaday, "Common Faults in the Mounting of Quadrupeds," *Third Annual Report of the Society of American Taxidermists* (1882–1883), p. 67.

21. Roland Barthes refers to the photograph as a "flat death" that presents the pastness of a certain moment while testifying to the camera click that "shot" the target. *Camera Lucida: Reflections on Photography*, trans. Richard Howard (New York: Hill and Wang, 1981), p. 92.

22. Oliver Wendell Holmes, "The Stereoscope and the Stereograph," *Atlantic Monthly*, June 1859, p. 748.

23. Thayer and Thayer, *Concealing Coloration in the Animal Kingdom*, p. 4.

24. He compares the preservation that results through the photograph with that enabled by amber, which preserves the bodies of insects for their arrival in the present in the museum "out of a distant past."

25. Barry Faulkner, *Sketches from an Artist's Life* (Dublin, NH: William Bauhen, 1973), pp. 43–44

26. Hornaday, *Taxidermy and Zoological Collecting*, p. 53.

27. In *On Photography*, Susan Sontag describes the connection between the photograph as evidence and the photograph as death. In Sontag's account, the very fact that the photograph provides proof that something happened-occurred-existed condemns it to death. In order to keep the evidence of the thing alive, the thing itself must remain dead. For Sontag, photographic death consists of a memorialization that renders permanent the fragility of a past spatiotemporal moment. On the other hand, this very control function condemns its object to death. She writes: "Just as the camera is a sublimation of the gun, to photograph someone is sublimated murder—a soft murder, appropriate to a sad, frightened

time.... All photographs are *memento mori*. To take a photograph is to participate in another person's (or thing's) mortality, vulnerability, mutability." Susan Sontag, *On Photography* (New York: Anchor Books: 1977), p. 15.

28. Abbott H. Thayer, "Further Remarks on the Law Which Underlies Protective Coloration," *The Auk* 13 (October 1896), p. 318.

29. Thayer's annotated copy of his April 1896 article offprint is held in the Ernst Mayr Library of the Museum of Comparative Zoology at Harvard University, MCZ Z-M, accessioned November 24, 1923, after p. 26.

30. F.S. Webster, "The Birth of Habitat Bird Groups," *Annals of the Carnegie Museum* 30 (1945), pp. 97–118; "Art in Aid of Science: Exhibition of Drawings, Paintings and Sculptures by Artists of the Museum of Natural History Showing Preparations for Groups and Figures," *New York Times*, May 12, 1907, p. SMA7.

31. Shell, "Skin Deep," pp. 91–121.

32. Abbott H. Thayer to Alfred R. Wallace, July 22, 1905, Alfred Russel Wallace Archive, Oxford University Museum of Natural History (OUMNH).

33. Faulkner, *Sketches from an Artist's Life*, p. 22.

34. Alfred R. Wallace to Abbott H. Thayer, July 11, 1898, Abbott H. Thayer Papers, Addition, AAA. Wallace mentions interest in Thayer's feather paintings.

35. Abbott H. Thayer to Alfred R. Wallace, July 22, 1905, Library of the Oxford University Museum of Natural History. Thanks to Stella Brecknell, librarian of the Hope Library.

36. *Ibid.*

37. *Abbott H. Thayer Exhibit Catalog*, Carnegie Institute, Pittsburgh, April 24–June 30, 1919, Abbott H. Thayer Papers; "Exhibit of Paintings and Studies by Abbott H. Thayer and Gerald H. Thayer, Brooklyn Museum, March 20–May 1, 1922," Cleveland Museum of Art Archives Ingalls Library Clipping Files, Thayer, Abbott, Abbott H. Thayer Exhibit of Protective Coloration Studies, Boston Museum of Fine Arts, March 1918; *Abbott H. Thayer Exhibit of Protective Coloration Studies*, Boston Museum of Fine Arts, March 1918. See also discussion of exhibitions in White, *Abbott H. Thayer*, p. 166.

38. Extended descriptions of feather and stencil paintings are included in the catalog for an exhibit at the Pratt Institute, Brooklyn, *Exhibition of Paintings Illustrating Protective Coloration in Nature by Abbott H. Thayer and Gerald H. Thayer*, Pratt Institute Art Gallery, 1924 (Pratt Institute Exhibit Catalogs, 1916–1924, reel 4859, AAA), entry 1.

39. Mary Fuertes Boynton, "Abbott Thayer and Natural History," *Osiris* 10.1 (1952), p. 543.

40. Abbott H. Thayer to Alfred R. Wallace, ca. February 1905. Alfred Russel Wallace Library, OUMNH.

41. Abbott H. Thayer to Alfred R. Wallace, July 22 1905, Library of the Oxford University Museum of Natural History.

42. *Exhibition of Paintings Illustrating Protective Coloration in Nature by Abbott H. Thayer and Gerald H. Thayer*, Pratt Institute Art Gallery, 1924 (Pratt Institute Exhibit Catalogs, 1916–1924, reel 4859, AAA), entry 2.

43. Ross Anderson, *Abbott Handerson Thayer*, exhibition catalogue, (Syracuse, NY: Everson Museum, 1982), p. 28. Barry Faulkner refers to the "well-worn quarto edition" that Thayer had since childhood kept close by his side. *Sketches from an Artist's Life*, pp. 19–22.

44. Thayer and Thayer, *Concealing Coloration in the Animal Kingdom*, p. ix.

45. *Ibid.*, p. 172.

46. *Ibid.*, frontispiece.

47. Louis Agassiz Fuertes, "Concealing Coloration in the Animal Kingdom," *Science*, n.s. 32.823 (October 7, 1910), pp. 466–69; see *Book Review Digest: Sixth Annual Cumulation of Book Reviews of 1910 in One Alphabet* (Minneapolis: H.W. Wilson, 1910), p. 392.

48. E.B. Titchener, "An Arraignment of the Theories of Mimicry and Warning Colors: Concealing Coloration in the Animal Kingdom: An Exposition of the Laws of Disguise through Color and Pattern, Being a Summary of Abbott H. Thayer's Discoveries," *The American Journal of Psychology* 21.3 (July 1910), pp. 500–504.

49. *Ibid.*, p. 503.

50. T.D.A. Cockerell, "Nature's Game of Hide and Seek: A Review of Concealing Coloration in the Animal Kingdom," *The Dial* (July 16, 1910), p. 33.

51. Roosevelt had traveled on an African safari sponsored by the Smithsonian Institution and the National Geographic Society and had returned to the United States with tons of skins and bones (11,397 items, total). "Roosevelt Specimens," *New York Times*, April 7, 1910, p. 4.

52. Theodore Roosevelt, "Revealing and Concealing Coloration in Birds and Mammals," *Bulletin of the American Museum of Natural History* 30 (1911), p. 230.

53. Thayer and Thayer, *Concealing Coloration in the Animal Kingdom*, p. 4.

54. Vivian Sobchack, "The Scene of the Screen: Envisioning Cinematic and

Electronic 'Presence," in Hans Ulrich Gumbrecht and K. Ludwig Pfeiffer (eds.), *Materialities of Communication*, trans. William Whobrey (Stanford: Stanford University Press, 1994), p. 92.

55. Roosevelt, "Revealing and Concealing Coloration in Birds and Mammals," pp. 227–28.

56. Roosevelt, *African Game Trails: An Account of the African Wanderings of an American Hunter-Naturalist* (New York: C. Scribner's Sons, 1910), p. 499.

57. Alexander Nemerov argues that the intensity of Roosevelt's ire was rooted in the period's more general crisis of manhood in light of advances in technology and shifts in sexual mores. Alexander Nemerov, "Vanishing Americans: Abbott Thayer, Theodore Roosevelt and the Attraction of Camouflage," pp. 50–81.

58. Roosevelt, "Revealing and Concealing," p. 226.

59. Roosevelt, *African Game Trails*, p. 491.

60. Thomas Barbour and John C. Phillips, "Concealing Coloration Again," *The Auk* 28 (April 1911), pp. 179–88. Abbott H. Thayer, "Concealing Coloration: A Demand for Investigation of My Tests of the Effacive Power of Patterns," *The Auk* 28 (October 1911), pp. 460–64. Francis H. Allen, "Remarks on the Case of Roosevelt vs. Thayer, with a Few Independent Suggestions on the Concealing Coloration Question," *The Auk* 29 (October 1912), pp. 489–507.

61. Frank Chapman, "The Scientific Value of Bird Photographs," Letter to *The Auk*, December 10, 1912, *The Auk* 30 (January 1913), pp. 147–49. Francis H. Allen, "The Concealing Coloration Question," Letter to *The Auk*, February 14, 1913, *The Auk* 30 (April 1913), pp. 311–17. Abbott H. Thayer, "Naturalists and Concealing Coloration," Letter to *The Auk*, September 11, 1913, *The Auk* 30 (October 1913), p. 618.

62. Nemerov, "Vanishing Americans," p. 56.

63. "Patterns and White," Abbott H. Thayer to Editors of *New York Sun*, ca. 1915. Abbott H. Thayer Papers, Addition, AAA. Abbott H. Thayer, "An Arraignment of the Theories of Mimicry and Warning Colors," *Popular Science Monthly* 75 (December 1909), pp. 550–70.

64. Abbott H. Thayer, "Concealing Coloration," *Popular Science Monthly* 79.1 (July 1911), p. 26.

65. Abbott H. Thayer, "Concealing Coloration: An Answer to Theodore Roosevelt," *Bulletin of the American Museum of Natural History* 31 (September 14, 1912), pp. 314–15.

66. *Ibid.*, pp. 315–16.

67. "Patterns and White," Abbott H. Thayer to editors of *New York Sun*, ca. 1915, Abbott H. Thayer Papers, Addition, AAA. See also mimeographs of letters that Thayer wrote to Franklin D. Roosevelt in 1914 and 1917 advising that boats should be painted white, not gray.

68. Abbott H. Thayer, "Teaching Britannia Her Job," *New York Tribune*, August 13, 1916, p. 4; Abbott H. Thayer, "Camouflage," *Scientific Monthly* 7.6 (December 1918), pp. 481–94.

69. Abbott H. Thayer, Letter to *Auk*, July 10, 1911, *The Auk* 28 (August, 1911), p. 146.

70. Alfred R. Wallace, "The Origin of Human Races and the Antiquity of Man Deduced From the Theory of 'Natural Selection,'" *Journal of the Anthropological Society of London* 2 (1864), p. 161.

71. *Ibid.*, p. 168.

72. *Ibid.*

73. Charles Darwin, *The Descent of Man and Selection in Relation to Sex: Volume One* (New York: D. Appleton and Co., 1872), p. 142.

74. Abbott Thayer to John Singer Sargent, ca. 1915, Abbott H. Thayer Papers, Addition, AAA.

75. Alexander Nemerov considers this idea in terms of a "failure to conform to ideals of heroic masculinity." Nemerov, "Vanishing Americans," p. 79.

76. William James, Jr. quoted by White, *Abbott H. Thayer*, p. 190.

77. White, *Abbott H. Thayer*, p. 51.

78. Hugh Cott, *Adaptive Coloration in Animals* (Oxford: Oxford University Press, 1941.)

79. In another letter, this one dated May 6, 1917, Faulkner wrote from New York City to his friend Harold Witter Bynner: "Since the first of March, I have been so busy trying to organize a body of artists to practice camouflage for the army. I also have some fine designs for uniforms that Abbott Thayer has made. Fry has been working with me and we have an association of about 200. We put the idea up to the government through the Naval Consulting Board, from them to the Council of National Defense and now the War College is going to appoint a committee to see if we're any good." Barry Faulkner Papers, Box 1, AAA.

80. Roy Behrens, *False Colors: Art, Design, and Modern Camouflage* (Dysart, IA: Bobolink Books, 2002), p. 63.

81. Director of Equipment Stores to John Singer Sargent, January 10, 1916, London, Abbott H. Thayer Papers, Addition, AAA; J. Stevens, British War Office to Abbott H. Thayer, August 14, 1916, London, Abbott H. Thayer Papers, Addition, AAA.

82. Behrens, *False Colors*, p. 63.

83. "Faking as an Art in Conducting War: Fifty American Artists Form an Association for Camouflage to Aid the Army," *New York Times*, June 24 1917, p. 17; *The Plattsburger* (New York: Wynkoop, Hallenbeck and Crawford, 1917).

84. Cott, *Adaptive Coloration in Animals*, p. 173

CHAPTER TWO: MENDING THE NET

1. Imperial War Museum (IWM) Photo, Album 055, Q95925-Q96060, *The British Army School of Camouflage 1917–1918*. Prints donated by Miss Florence Mottershead, 30 Ladies Mile Road, Brighton BN1 8QF. The inscription continues: "The entire contents of this book have been taken or photographed by the members of the photographic section attached to the Camouflage School R. E. stationed in Kensington Gardens, London for two years and transferred to 21 Camp, Larkhill, Salisbury Plain, June 1919.... Under the title 'camouflage,' from a French slang term, having much the same meaning as the English word 'faking,' the study of concealment was instituted first by the French, and later by the British."

2. IWM Photo Q95947.

3. IWM Photo Q95943.

4. *La grande illusion*, Jean Renoir's 1937 film about World War I, concerns life among a group of diverse captured French soldiers in a prisoner of war camp. In one scene, Rosenthal, one of the film's two protagonists, receives a pair of ladies stockings from home and dances around, stretching them over his head while chanting "les bas, les bas."

5. E. Torday and T. A. Joyce, *Notes ethnographiques sur les peuples communément appleés Bakuba* (1910), cited in Mary Douglas, "Raffia Production in the Lele Economy," *Africa: Journal of the African Institute* 28.2 (April 1958), p. 109.

6. Raymond Myerscough-Walker, "Camoufleur and his Craft part 2," *The Builder* 157 (October 1939), p. 457.

7. Michel Foucault, "Panopticism," in *Discipline and Punish: The Birth of the Prison*, trans. Alan Sheridan (New York: Vintage, 1995), p. 195–228.

8. Lieutenant Colonel Frederick J. Hutchison and Captain H. G. MacGregor,

Military Handbook: Military Sketching and Reconnaissance (London: C. K. Paul, 1878), p. 71; William Willoughby and Cole Verner, *Rapid Field-Sketching and Reconnaissance* (London: W. H. Allen, 1889).

9. Peter Mead, *The Eye in the Air: History of Air Observation and Reconnaissance for the Army, 1785–1945* (London: Her Majesty's Stationery Office, 1983).

10. Captain R. S. S. Baden-Powell, *Reconnaissance and Scouting: A Practical Course of Instruction, in Twenty Plain Lessons, for Officers, Non-Commissioned Officers, and Men* (London: W. Clowes and Sons, 1884). Baden-Powell went on to become a lieutenant general, as well as to publish *Scouting for Boys* in 1908, which became the official *Boy Scout Handbook*, a version of which is still in use today.

11. H. J. McKenney, *Scout's Handbook and Instructor* (Kansas City, MO: Franklin Hudson, 1912).

12. As in the Magritte painting *Ceci n'est pas une pipe.*

13. Paul Virilio, *La machine de vision* (Paris: Galilée, 1988), p. 38, quoted by Martin Jay in *Downcast Eyes: The Denigration of Vision in Twentieth-Century French Thought* (Berkeley: University of California Press, 1994), p. 211; Cecile Coutin "Le camouflage pendant la Première Guerre mondiale I," *Historiens-Géographes* no. 321 (December 1988), pp. 265–82.

14. David Henderson, *The Art of Reconnaissance* (London: John Murray, 1914), p. 173.

15. *Ibid.*

16. Along these lines, see *Notes for Observers*, issued July 26, 1915 and carried by order in all Royal Flying Corps (RFC) airplanes. The system of visualization and notation employed therin was rooted in the conventional signs of the nineteenth century. Specific uses for *Notes for Observers*, used by members of the RFC attached to the Expeditionary Force General Headquarters 1149, were described in a letter from the field to Headquarters (Expeditionary Force G.H.Q 1149), June 23, 1915. National Archives (NA): Public Records Office (PRO) AIR/1/128/15/40/175. The AIR series contains of records of the British Air Ministry Files; this subseries holds records from the Air Historical Branch.

17. Felix Tournachon (Nadar) made the first aerial photograph from a tethered balloon in 1858 and went on to produce several Paris city series in the 1860s. Nadar suggested military applications almost immediately, and limited balloon field tests were made during the American Civil War. Beaumont Newhall, *The History of Photography from 1839 to the Present* (New York: Museum of Modern Art, 1982), p. 110.

18. Major Geoffrey Toye, *History of Photography: Being a General History of Air Photography 1914–1918, Its Camera, Lens Etc.* NA: PRO AIR 1/724/91/6/1.

19. Herbert E. Ives, *Airplane Photography* (Philadelphia: J. P. Lippincott, 1920), p. 15.

20. Squadron #27 RFC Reports of Photographic Reconnaissance Reports 1918 January–1918 March, PRO AIR 1/1384/204/25/7; also Caleb Arnold Slade Papers, pp. 1–3, Archives of American Art (AAA); and Memoir of Lieutenant Colonel H. Wyllie, #23 Squadron, entry for June 8, 1916, third reel in the collection Diaries from the Somme, assembled by the Imperial War Museum, accessed on microfilm at the library of the Institute for Historical Research at the University of London.

21. In-depth accounts of the emergence of British aerial photographic surveillance during World War I include Peter Mead, *The Eye in the Air: A History of Air Observation and Reconnaissance for the Army 1785–1945* (London: Her Majesty's Stationery Office, 1983); Ursula Powys-Lybbe, *The Eye of Intelligence* (London: William Kimber and Co, 1983); and Roy Conyers Nesbit, *Eyes of the RAF: A History of Photo-Reconnaissance* (London: Allan Sutton Limited, 1996).

22. Toye, *History of Photography*, PRO AIR 1/724/91/5.

23. John Keegan, *The Face of Battle: A Study of Agincourt, Waterloo and the Somme* (London: Penguin, 1976); see also Toye, *History of Photography*, p. 6.

24. *Elementary Training of Pilots and Observers in the use of Aerial Cameras, 6/9/17–19/9/17*, PRO AIR 1/664/17/122/697.

25. J. T. C. Moore- (Air Ministry) to Charles Fairbairn (RAF), October 11, 1918, in "Reminiscences of J. T. C. Brabazon," PRO AIR 1/724/91/5.

26. Toye, *History of Photography*, p. 8.

27. However, "in the lack of time a *memory sketch* can be made." Lieutenant Colonel Frederick J. Hutchison and Captain H. G. MacGregor, *Military Handbook: Military Sketching and Reconnaissance* (London: C. K. Paul, 1878), p. 66. See also Toye, *History of Photography*, p. 5.

28. Alan Sekula, "The Instrumental Image: Steichen at War," *Artforum* 14.4 (December 1975), p. 28.

29. Lorraine Daston and Peter Galison, "Mechanical Objectivity," in *Objectivity* (New York: Zone Books, 2007), pp. 115–90; Peter Galison, "Judgment against Objectivity," in Caroline Jones and Peter Galison (eds.), *Picturing Science, Producing Art* (London: Routledge, 1998), pp. 327–60.

30. Sekula, "The Instrumental Image: Steichen at War," p. 30; also Paul K.

St. Amour, "Modernist Reconnaissance," *Modernism/Modernity* 10.2 (April 2003), pp. 349–80.

31. In a sense, the aerial photograph might be understood as a counterpoint to turn of the century *trompe l'oeil* painting. Johanna Drucker, "Harnett, Haberle, and Peto: Visuality and Artifice among the Proto-Modern Americans," *Art Bulletin* 74,1 (March 1992), pp. 37–50.

32. Anita V. Mozley, *Eadweard Muybridge: The Stanford Years* (Stanford, CA: Board of Trustees of the Leland Stanford Junior University, 1972), p. 19; also, Robert Bartlettt Haas, *Muybridge: Man in Motion* (Berkeley: University of California Press, 1976) pp. 160–62.

33. Eadweard Muybridge, preface to *The Complete Works: An Electro-Photographic Investigation of Consecutive Phases of Animal Motion—781 Plates* (Philadelphia: University of Pennsylvania, 1887), n.p.

34. Straight lines—always perpendicular to the black bars—split ground from background, creating two rectangular subdivisions, the upper of which is twice the area of the lower. Each series of photographs breaks into congruent rectangular units of five by nine centimeters.

35. Photographs of Tirpitz, Study of RNAS Photographs nos. 1103 to 1106 inclusive of Tirpitz Battery, by No. 2. Squadron. No. 32 in RNAS Studies, No. 1 to 50, April–July 1917, PRO AIR 1/81/15/9/201.

36. Sekula, "The Instrumental Image: Steichen at War," p. 28.

37. As St. Amour has argued, "the stereoscope fused the distinct relief displacement of its stereopairs into an apparent three-dimensionality." St. Amour, "Modernist Reconnaissance," p. 363.

38. Solomon J. Solomon, "Visual Deception in Warfare," April 5–20, 1916, PRO AIR 1/530/16/12/89, p. 3.

39. Photographs of Tirpitz, Study of RNAS Photographs, PRO AIR 1/81/15/9/201.

40. Olga Somech Phillips, "Lt. Colonel Solomon—Camoufleur," in *Solomon J. Solomon: Memoir of Peace and War* (London: Herbert Joseph, 1933), p. 118.

41. Solomon J. Solomon, "A Method by which the Round Object Can be Reduced to the Flat," in *The Practice of Oil Painting and Drawing as Associated with It* (London: Speeley and Co., 1911, reprinted 1914), p. 21–23.

42. Solomon J. Solomon, *Visual Deception in Warfare*, April 1916, PRO AIR 1/530/16/12/89, p. 1. For an analysis thereof, see "Notes on Colonel Solomon's

Ideas on German Camouflage. March 30 1918," and unsigned letter mailed to Charles Fairbairn December 6, 1919. PRO AIR 1/1/4/21.

43. Solomon, *Visual Deception*, p. 2.

44. Guy Hartcup, *Camouflage: A History of Concealment and Deception in War* (New York: Scribners, 1980), pp. 171–72. Also C. H. R. Chesney, The Art of Camouflage (London: Robert Hale, 1941), pp. 66–68.

45. Details on the design and construction of the first operation post tree by the British are found in the personal notebook of G. C. Leon Underwood, consisting of paintings, sketches and photographic records. Imperial War Museum Art Department.

46. Oliver Percy Bernard, *Cock Sparrow: A True Chronicle* (London: Jonathan Cape, 1936), p. 205. (André Mare, the Frenchman and cubist, assisted in the first British O.P. tree constructions.)

47. Solomon's postwar analysis of suspected German camouflage measures appeared in 1920. Solomon J. Solomon, *Strategic Camouflage* (London: J. Murray, 1920).

48. Roy Behrens, *False Colors: Art, Design and Modern Camouflage* (Iowa City: Bobolink Books, 2002), p. 75; Coutin, "Le camouflage pendant la Première Guerre mondiale I," p. 265.

49. C. H. R. Chesney, *The Art of Camouflage* (London: Robert Hale, 1941), p. 66.

50. Adrian Cornwell-Clyne, "Notes on Concealment, 1915" (twelve-page manuscript), Private Papers of A. Cornwell-Clyne, Imperial War Museum Documents (IWM DOC) 86/53/1, p. 7. Adrian Klein, son of the famous music critic Hermann Klein, changed his name to Cornwell-Clyne after the war so as to distance himself from his Jewish roots. With his new name, he went on to become an important inventor and developer of new color and 3D filmmaking processes in the 1920s and 1930s.

51. Chesney, *The Art of Camouflage*, p. 66.

52. Nicholas Rankin, *Churchill's Wizards: The British Genius for Deception 1914–1945* (London: Faber and Faber, 2008), p. 119–22.

53. Royal Engineers Special Works Park, March 1916–July 1918, PRO WO 95/120; Royal Engineers, Southern Special Works Park, February 1917–April 1919, and Northern Special Works Park, February 1917–May 1918," PRO WO 95/4058; *Camouflage Experimental Establishment*, PRO T1/12557. The PRO WO series

consists of the British War Office files; the T1 series contains Treasury records.

54. Certainly, some factors leading to the emergence of strategic and systematic concealment practices by the modern military had less to do with vertical photography, and more to do with weapons technology—the emergence of long-range artillery—and its transformation of horizontal battle starting in the later nineteenth century. As guns fired farther and with greater accuracy, hiding became more and more critical and concealment more generally practiced. To cite one well-known example, the British Army switched its uniforms to khaki and khaki serge by 1902, motivated primarily by the need for concealment from ground detection. But in point of fact, such changes in weapons technologies cannot be disentangled in any significant way from changes in reconnaissance technologies; these new long-range firing capabilities both prompted, and were accelerated by, the development of military airplanes and aerial photographic surveillance.

55. Francis Wyatt, s.v. "Military Camouflage," *Encyclopedia Britannica*, 11th ed. (1922).

56. "Minute du 24 octobre 1915. Gouvernment Militaire de Paris," SHAT 7N410, cited by Coutin, "Le camouflage pendant la Première Guerre mondiale I," p. 270.

57. See Solomon, *Strategic Camouflage*. Patrick Wright discusses Solomon's (mis)reading of the German camouflage schemes of World War I in *Iron Curtain: From Stage to Cold War* (Oxford: Oxford University Press, 2007), pp. 143–50.

58. Coutin, "Le camouflage pendant la Première Guerre mondiale I," p. 63.

59. Exact employee numbers are tabulated in charts within Special Works Park, March 1916–July 1918, PRO/WO 95/120; Southern Special Works Park, February 1917–April 1919, and Northern Special Works Park, February 1917–May 1918, PRO WO 95/4058.

60. Tyler Stovall, "Color-Blind France?: Colonial Workers during the First World War," *Race & Class* 35.2 (October 1993), pp. 35–55.

61. Chesney, *The Art of Camouflage*, p. 64.

62. *Ibid.*, p. 71.

63. Caleb Arnold Slade, "Notes on Camouflage," pp. 2 and 2.5, C. Arnold Slade Papers, AAA.

64. Toye, *History of Photography*, p. 8.

65. Nicholas Reeves, *Official British Film Propaganda During the First World War* (London: Croon Helm, 1986), p. 148.

66. Solomon, *Visual Deception in Warfare*, pp. 3–4.

67. Chesney, *The Art of Camouflage*, p. 78.

68. *Ibid.*, p. 76.

69. *Ibid.*, p. 79.

70. *Ibid.*, p. 84.

71. *Ibid.*, p. 16.

72. *Ibid.*, p. 79.

73. *Ibid.*, p. 78.

74. General François de Paule Michel Jacques Raymond de Fossa, *Conférence sur le camouflage faite par le Chef d'escadron de Fossa*, Cours d'Études de Vitry-Le-François (n.p., 1917), and *Instructions sur le camouflage* (Paris: Imprimerie Adrien Maréchal, 1918).

75. By 1917, the U.S. Army was distributing standardized documents about camouflage to troops going to the front lines, for example: *Notes on Camouflage* (Washington D.C.: Government Printing Office and Army War College, 1917) and *Camouflage for Troops of the Line* (Washington, D.C.: Government Printing Office and Army War College, 1918). The War Department issued the latter document on January 3, 1918, and the French and British Governments issued similar manuals throughout 1918.

76. Homer Saint-Gaudens, "Camouflage and Art," *The Art Bulletin* 2.1 (September 1919), p. 29.

77. Jean-Louis Commolli, "Machines of the Visible," in Stephen Heath and Teresa de Lauretis (eds.), *The Cinematic Apparatus* (London: Macmillan, 1980), p. 132.

78. Slade, "Notes on Concealment," p. 3.5.

79. Hew Strachan, "Training, Morale and Modern War," *Journal of Contemporary History* 41.2 (2006), pp. 223–25.

CHAPTER THREE: HOW NOT TO BE SEEN

1. *Monty Python's Flying Circus*, episode 24, first aired December 8, 1970 and included a preliminary version of "How Not to Be Seen" (see http://video.google.com/videoplay?docid=1111770357062149413#), which was then refilmed for incorporation into Ian MacNaughton's Monty Python compendium *And Now for Something Totally Different* (1971), It is this second version that ends with Cleese, the narrator, sitting in the field, awaiting his own death, as it turns out. See http://

www.youtube.com/watch?v=uoWOIwlXE9g.

2. The amateur film *Formation of the Home Guard, Thornton Bradford* (1944), directed by Harold Whitehead and the Bradford Cine Club, documents standard training practices and military exercises of the unit between 1939 and 1944, as well as the shift from the LDV to the Home Guard structure including training in marksmanship and camouflage. (Yorkshire Film Archive 1214.)

3. Surrealist artist and camouflage instructor Roland Penrose (knighted in 1996) used media to educate the masses in defensive subterfuge techniques in his widely circulated booklet *Home Guard Manual of Camouflage* (London: G. Routledge, 1941). For the larger context, see "Prime Minister on the Home Guard" and "Object of Home Guard" in John Langdon Davies (ed.), *The Home Guard Training Manual* (London: Wyman & Sons, 1940), pp. 8 and 10–20.

4. S. P. MacKenzie, *The Home Guard: A Military and Political History* (Oxford: Oxford University Press, 1995), pp. 21–27. On the mobile cinema units, see Paul Swann, *The British Documentary Film Movement, 1926–1946* (Cambridge: Cambridge University Press, 1989), pp. 169–70.

5. One especially notable example of a such a film was *Went the Day Well?* (1942). It was based on a short story by Graham Greene about an English village overtaken by Nazi paratroopers disguised as British soldiers. The film's director was Alberto Cavalcanti, the Brazilian-born documentary filmmaker and producer who later commended Len Lye on his works of realist wartime nonfiction *Kill or Be Killed* in particular. Jo Fox, *Film Propaganda in Britain and Nazi Germany: World War II Cinema* (Oxford: Berg, 2007), pp. 152–56. On the various forms of wartime film sponsorship, see Francis Thorpe and Nicholas Pronay, *British Official Films in the Second World War: A Descriptive Catalogue* (Oxford: Clio Press, 1980), pp. 40–48.

6. Thomas M. Pryor, "The Shorts Have It: Being a Resume of Five Newly Arrived Topical Film Subjects from Britain," *New York Times*, August 8, 1943, p. X3.

7. Shooting script and press materials for *Kill or Be Killed*, held among the records of the Film and Television Division of the Central Office of Information, the post World War II incarnation of the MOI. National Archives: Public Records Office (PRO) INF 6/479. (INF designates the series of papers of the Central Office on Information.)

8. "Review of *Kill or Be Killed*," *Documentary News Letter*, January 1943, p. 165; "Letter to the Editor," *Documentary News Letter*, February 1943, p. 184.

9. Thomas M. Pryor, "British Fact Films Find Wide Audience Here," *New York Times*, January 9, 1944, p. X3.

10. Jürgen Berger, "Listen to Britain: Strukturen und Arbeitsweisen der Films Division des Ministry of Information 1939–45. Ein Beitrag zur Administrations- und Produktionsgeschichte britischer Filmpropaganda," Ph.D. diss., University of Konstanz, 2001, p. 186.

11. "Discussion with the Ministry of Information on Educational Films 1943–1945," in *Catalog of Films Made and Acquired by the Ministry of Information, from July 1st till December 31st, 1942*, p. 192. PRO ED 121/594. (ED designates the series of Records of the Department of Education and Science of the Board of Education in the National Archives Public Records Office.)

12. Pryor, "The Shorts Have It," p. x3.

13. "Review of Kill or Be Killed," p. 165.

14. "Camouflage," *New York Times Magazine*, May 25, 1941, p. 14.

15. Photographic film was an increasingly omnipresent means of surveillance, a medium from which one sought to efface oneself at the same time as it was hoped to be the most practical media for instruction in how to do so.

16. The path from the sensibility and aesthetic of 1942 to the media environment in which we live today is not by any means direct. To the best of my knowledge, game programmers in 2011 are not watching World War II training films, nor are gamers looking to 1930s experimental animation for inspiration. But we are living in a world anticipated by Lye's style.

17. Alberto Cavalcanti, "Presenting Len Lye," *Sight and Sound* (Winter 1947–1948), p. 135.

18. Valuable sources on Lye's life and artistic career include Roger Horrocks's *Len Lye: A Biography* (Auckland: Auckland University Press, 2001). See also Horrocks's essay "Len Lye—Origin of His Art" in the exhibition catalog he coedited with Jean-Michel Bouhours, *Len Lye* (Paris: Editions du Centre Pompidou, 2000), pp. 178–83. Sarah Davy's very short essay "Lye: the Film Artist in Wartime" in the same exhibition catalog, pp. 196–98, outlines Lye's film output with the Realist Film Unit between 1941 and 1944.

19. Tyrus Miller, "Documentary/Modernism: Convergence and Complementarity in the 1930s," *Modernism/Modernity* 9.2 (April 2002), pp. 226–41.

20. Len Lye, quoted in Ray Thorburn, "Interview with Len Lye," *Art International* 19 (April 1975), p. 65.

21. In the animal kingdom, furred or feathered skin has often evolved so as either to make an organism stand out against its background or to blend in visually with that same environment. In this immersive sense, a skin acts so as to visually collapse the visible boundary between the organism and its environment. Thayer had argued human tattooing practices have at times served as analogous function. (See Chapter 1.)

22. Len Lye, quoted Adrienne Mancia and Willard Van Dyke, "The Artist as Filmmaker: Len Lye," *Art in America* 54 (July–August 1966), p. 106.

23. Simon Brown, "Dufaycolor—The Spectacle of Reality and British National Cinema" (London: n.p., 2004); Adrian Cornwell-Clyne, *Colour Cinematography*, 2nd ed. (London: Chapman and Hall, 1939).

24. Records on Len Lye's *Trade Tattoo* (abstract color film, working title "Post Early," GPO Film Unit, 1937), PRO INF 6/311.

25. The *Oxford English Dictionary* credits Captain Cook with the first English usage of this term. In Cook's travels in Polynesia and Tahiti, the very regions from which Lye took much of his inspiration, he noted the body-art practices. In his 1769 account, he begins, "This method of Tattowing, I shall now describe." James Cook, *Captain's Cook's Journal during the First Voyage Round the World* (Whitefish, MT: Kessinger, 2004), p. 159.

26. William C. Wees, *Recycled Images: The Art and Politics of Found Footage Films* (New York: Anthology Film Archives, 1993), p. 26.

27. Lye's interest in New Zealand textiles and Polynesian tapa and body-tattooing cloth had included the production of batiks modeled on his observations of Samoan tapa bark patterns, abstract renderings of plants and animals. Horrocks, *Len Lye*, pp. 64–65 and 85–90.

28. Cornwell-Clyne, "Notes on Concealment," 1915, twelve-page manuscript, A. Cornwell-Clyne Collection, Imperial War Museum Department of Documents, IWM DOC 86/53/1, p. 2.

29. Roger Horrocks, *Len Lye: A Biography* (Auckland: New Zealand University Press, 2001), p. 198.

30. Ministry of Information officer to Robert Wood, London, December 30, 1943, p. 1, in *Discussion with M.O.I. on Education Film, 1943–1945*, PRO ED 121/594.

31. *History and Development of the Camouflage Section and S.O.E, 1941–1945*, in Special Operations Executive: Histories and War Diaries: Registered Files. PRO HS 7/49. (HS designates the series of records of the Special Operations Executive

in the National Archives Public Records Office.)

32. N. A. D. Armstrong, *Fieldcraft and Stealth* (Aldershot, UK: Gale and Polden, 1942), p. 11.

33. *Ibid.*

34. *Ibid.*, pp. 18–19.

35. Ion Idriess, *Sniping* (London: Angus and Robertson, 1942), p. 38.

36. Gerald Thayer and Abbott H. Thayer, *Concealing Coloration in the Animal Kingdom: An Exposition of the Laws of Disguise through Color and Pattern, Being a Summary of Abbott H. Thayer's Discoveries* (New York: Macmillan, 1909), p. 5.

37. "Fieldcraft and Tactics for the Royal Air Force Regiment (based on the teaching of the School of Infantry), 1st Edition," PRO AIR 10/3723.

38. Arts Enquiry, *The Factual Film: A Survey* (London: Oxford University Press, 1947), pp. 71–74. Also, PRO ED 121/594.

39. For example, *Fieldcraft and Tactics for the Royal Air Force Regiment: Promulgated for the Information and Guidance of all Concerned: Restricted*, April 1944, PRO AIR 10/3723, p. 1.

40. *Camouflage and Fieldcraft Part I: Fieldcraft and Offensive Planning*, 2 reels, 1942, produced by the Army Kinematograph Service (AKS) as part of the AKS Battle Drill Series. Held at the Imperial War Museum (IWM) Film Collection SKC 22. Other films in the series include *Camouflage and Fieldcraft: Prepare for Battle* (IWM Film WOY 266), and *Camouflage and Fieldcraft: Movement* (IWM Film SKC 23).

41. The slide show titled *Individual Concealment*, for example, is a set of twenty-six glass lantern slides produced by the Camouflage Development and Training Center, part of the British Army, released at the same time as Lye's *Kill or Be Killed*. The slide show comes with a specific text to be read by an infantry training officer. The *Individual Concealment* show both begins and ends with allusion to the animal world and advises students that they must constantly see themselves as if from the enemy's perspective. *Individual Concealment*, Box of Glass Lantern Slides, Imperial War Museum Department of Documents, Papers of Havinden A E Captain OBE, 74/165/8 & 8A–C.

42. James Chapman, *The British at War: Cinema, State and Propaganda 1939–1945* (London: I. B. Tauris, 1998), pp. 86–114.

43. Basil Wright to Len Lye, London, November 11, 1942, re: *Kill or Be Killed*, microfiche 201, Len Lye Collection, Govett-Brewster Art Gallery.

44. Robert Manvell, "They Laugh at Realism," *Documentary Educational New Letter* (March 1943), p. 188.

45. Ion Idriess, *The Scout* (London: Angus and Robertson, 1943), p. 66.

46. Alexander R. Galloway, *Gaming: Essays on Algorithmic Culture* (Minneapolis: University of Minnesota Press, 2006), p. 41.

47. Edward Branigan, *Point of View in the Cinema: A Theory of Narration and Subjectivity in Classical Film* (Berlin: Mouton Publishers), p. 80.

48. Roger Caillois, "Mimicry and Legendary Psychasthenia," (1935), trans. John Shepley, *October* 31 (Winter 1984), p. 31.

49. *Clothes and the Man* (1941), Analysis Film Limited for the Royal Air Force, produced by IWM Film AMY 134.

50. "Review of *Kill or Be Killed*," *Documentary News Letter* (January 1943), p. 165.

51. Tim Harrison Place, "Lionel Wigram, Battle Drill and the British Army in the Second World War," *War in History* 7.4 (2000), p. 445.

52. This style of fieldcraft training film continued to be made and remade, as in *Individual Fieldcraft* (1965) (IWM Film DRA 1201), *The Snipers* (1971) (IWM Film DRA 1285), and *Fieldcraft* (1990) (IWM Film DRA 1725).

53. Pryor, "The Shorts Have It," p. X3.

54. Cavalcanti, "Presenting Len Lye," p. 136.

55. Shooting script and press materials for *Kill or Be Killed*, PRO INF 6/479.

56. Idriess, *Sniping*, p. 8.

57. A precise count of how many people, and of what background, watched *Kill or Be Killed* during World War II is outside the purview here. Except for unreliable surveys done of particular World War II films (not including *Kill or Be Killed*) produced by the Ministry of Information and not necessarily trustworthy, data are not available.

58. Jean Lababut, "Last Lecture, February 4, 1944," Box 52, Jean Labatut Papers CO709. Used by permission of Princeton University Libraries.

59. *Ibid.* "In any case we still have to acquire a fully developed concept of camouflage.... Nature, history and Art are three sources of information, three experimental laboratories which with continuous scientific research can help us to acquire a fully-developed concept of camouflage."

60. *Ibid.*

61. Labatut, "Lecture One Notes," p. 1, Jean Labatut Papers.

62. Len Lye, "Experiment in Colour," *World Film News* (December 1936), p. 33.

63. Idriess, *Sniping*, p. 13.

64. Kaja Silverman, *The Threshold of the Visible World* (New York: Routledge, 1996), p. 133.

65. The lines and stencil motifs in Lye's animation work can be interpreted in a similar way. He crossed the boundaries of the film "frame," exceeding what might possibly have been captured by a cinema camera.

66. Caillois, "Mimicry and Legendary Psychasthenia," p. 31.

67. Evocative of the process of cinematic "suture" first laid out by Jean-Pierre Oudart in "Cinema and Suture," trans. Kari Hanet, *Screen* 18.4 (Winter 1978), pp. 35–47.

68. Idriess, *Sniping*, p. 38.

69. Mancia and Van Dyke, "The Artist as Filmmaker: Len Lye," p. 106.

70. *Ibid.*, p. 102.

71. "Review of Kill or Be Killed," p. 165.

72. Tim Lenoir, "All but War Is Simulation: The Military-Industrial Complex" *Configurations* 8.3 (2000), pp. 289–335.

73. Galloway provides a historical account and theoretical analysis of the emergence of the FPS game genre in "Origins of the First-Person Shooter" in *Gaming: Essays on Algorithmic Culture*, pp. 39–69.

74. *Ibid.*, p. 69.

75. First Person Shooter Strategies, "Movement and Camouflage," available at http://afirstpersonshooter.tripod.com/id8.html.

76. Idriess, *Sniping*, p. 13.

77. Mark Claypool, Kajal Claypool, and Feissal Damaa, "The Effect of Frame Rate and Resolution on Users Playing First Person Shooting Games," *Proceedings ACM/SPIE Multimedia Computing and Networking (MMCN) Conference* (San Jose, CA), January 2006, available at http://web.cs.wpi.edu/~claypool/papers/fr-rez.

78. See http://www.youtube.com/watch?v=af-LGNmqNAY. It was reposted in 2009 by [-A-] Washburn himself, a member of the Apostles BF2 Clan of Denmark, to his own channel on September 28, 2009, as "a tribute to Monty Python." Seehttp://www.youtube.com/watch?v=ZSK-BVsyK50. Washburn's clan membership profile is available at http://apps.decl.dev.ggl.com/personinfo.php?pid=3817796A.

CHAPTER FOUR: SUBJECT TO CHANGE

1. Philip K. Dick, *Philip K. Dick: Five Novels of the 1960s & 70s*, ed. Jonathan Lethem (New York: Literary Classics, 2008), p. 877. The novel was first published in 1977 by Vintage Books.

2. A precise explanation of the meaning of chameleonic or "active camouflage" within a military context, as well as the technological limitations surrounding its implementation, is provided in Kent W. McKee and David W. Tack, "Active Camouflage for Infantry Headwear Applications," Working Paper prepared by the Human*systems*® Incorporated/DRDC-Toronto Scientific Authority on behalf of the Canadian Department of National Defense, released February 2007, pp. 1–3 and 14–17.

3. Roger Caillois, "Mimicry and Legendary Psychasthenia," (1935), trans. John Shepley, *October* 31 (Winter 1984), p. 31.

4. In the film, cognitive dissonance, even dissociation, is evoked by the animation schema, as much as the story itself. Manohla Dargis remarks on this cognitive dissonance and its amplification though the animation process in her July 2006 film review. Manohla Dargis, "Undercover and Flying High on a Paranoid Head Trip," *New York Times*, July 7, 2006, p. E10.

5. Or to "out-Zelig" Zelig. Woody Allen's 1983 mockumenary *Zelig* (starring Allen and Mia Farrow) concerns an enigmatic human chameleon whose identity shifts so as to become the character of whomever is near him. Ultimately, his chameleonic nature proves psychologically debilitating, although he becomes an international phenomenon in the process.

6. Poulton continued, "Analogous powers exist in certain Crustacea and Cephalopoda," s.v. "Colours of Animals," *Encyclopedia Britannica*, 11th ed. On the significance of chameleonic color change in nature, see also James B. Murphy, *Chameleons: Johann von Fischer and Other Perspectives* (Salt Lake City: Society for the Study of Amphibians and Reptiles, 2005); Petr Nečas, *Chameleons: Nature's Hidden Jewels* (Frankfurt am Main: Edition Chimaira, 1999); and James Martin, *Masters of Disguise: A Natural History of Chameleons* (New York: Facts on File, 1992).

7. Charles Haskins Townsend, "Chameleons of the Sea: Some New Observations on Instantaneous Color Changes Among Fishes," *The Century*, September 1910, reprinted in the same year by The New York Zoological Society, p. 1.

8. *Ibid.*, p. 2.

9. "Attempts to photograph their normal changes in the large exhibition tanks

were, owing, on top of everything else, to the lack of light, seldom successful." *Ibid.*

10. Tom Gunning, "The Cinema of Attraction: Early Film, Its Spectator, and the Avant-Garde," in Robert Stam and Toby Miller (eds.), *Film and Theory: An Anthology* (Malden, MA: Blackwell, 2000), pp. 229–35. The essay was first published in *Wide Angle* in 1986.

11. The projector synthesized dynamic motion by literally propelling serial photographic frames into a kinetic, temporal progression—change in real time was enacted through movement.

12. Despite providing the phenomena's namesake, chameleons' capacity for "chameleonic" color change remains contested. Whereas scientists have long agreed upon the concealing functions of many of the rapid color changes of fishes and cephalopods, ironically, evolutionary biologists continue to quibble about whether the chameleon's color change actually *does* ever help the animal conceal itself from either predator or prey. Even the chameleon may be no chameleon at all. Unless otherwise noted, I use the adjective "chameleonic" to refer to the general phenomenon of rapid adjustable protective coloration, rather than to the specific group of animals by the same name in which it may or may not be found.

13. F. B. Sumner, "Does '"Protective Coloration"' Protect?: Results of Some Experiments with Fishes and Birds, *Proceedings of the National Academy of Sciences of the United States of America* 20.10 (October 15, 1934), p. 559.

14. Erasmus Darwin, *Zoonomia, or the Laws of Organic Life: Volume 1* (1794; Philadelphia: Edward Earle, 1818), p. 403. This passage is referred to by Mary A. Evans in "Mimicry and the Darwinian Heritage," *Journal of the History of Ideas* 26.2 (April–June 1965), p. 212.

15. Charles Darwin to J. S. Henslow, Rio de Janeiro, 18 May 1832, in *The Correspondence of Charles Darwin Volume 1, 1821–1836* (Cambridge: Cambridge University Press, 1985), p. 237. I thank Janet Browne for pointing me in the direction of this material.

16. Edward Poulton, "The Place of Mimicry in a Scheme of Defensive Coloration," *Essays on Evolution: 1889–1907* (Oxford: Oxford University Press, 1908), p. 304.

17. Charles Darwin, *On the Origin of Species by Means of Natural Selection, or The Preservation of Favoured Races in the Struggle for Life* (London: John Murray, 1859), pp. 72–73.

18. G. Sangiovanni, "Descrizione di un particolare sistema di organi cromo-foro espansivo-dermoideo e dei fenomeni che esso produce, scoperto nei molluschi cefaloso," *Enciclopedico Napoli* 9 (1819), pp. 1–13; Ernst Brücke, "Untersuchungen uber den Farbenwechsel des afrikanischen Chamaleons," *Denkschriften der Kaiserlichen Akademie der Wissenschaften, Mathematisch-Naturwissenschaftliche Klasse- Erste Abtheilung, Abhandlungen von Mitgliedern der Akademie* 4.1 (1852), pp. 179–210; and G. H. Parker, *Color Changes of Animals in Relation to Nervous Activity* (Philadelphia: University of Pennsylvania Pres, 1936), pp. 1–12.

19. Edward Bagnall Poulton, "Appendix A: Some Observations on Color Change among South African Chameleons," in Gerald Thayer and Abbott H. Thayer, *Concealing Coloration in the Animal Kingdom: An Exposition of the Laws of Disguise through Color and Pattern, Being a Summary of Abbott H. Thayer's Discoveries* (New York: Macmillan, 1909), p. 241.

20. "Mr. Thayer first suggested that the relative shades of the dark back, lighter sides and white undersides of animals were such as just to counterbalance the diminution of natural illumination from an open sky as we pass from the back down the sides to the under surface; that the object of this countergrading was to neutralize the shadow which would otherwise render the animal conspicuous. *C. pumilis* manifests the same principle in a dynamic form. The side that happens to be turned away from the light is brightened sufficiently to neutralize the shadow; the high illumination on the other side is toned down by darkening, the effect being that all appearance of solidity is dissipated. But for this adjustable countergrading, the varying degrees of illumination on the side and dorsal slope turned towards the light, combined with the strong shadow on the other side, would cause it to stand out among the leaves as an object of conspicuous solidity and thickness." Poulton, "Appendix A," pp. 242–43.

21. Edward Bagnall Poulton, *The Colours of Animals: Their Meaning and Use, Especially Considered in the Case of Insects* (London: Kegan Paul, Trench, Trübner, 1890), p. 732. Here Poulton based his explanation on Italian and German functional morphologists and endocrine scientists, including Sangiovanni and Brücke, who had investigated the function of the so-called "chromatophore" (or color changing) organ.

22. G. H. Parker, "Methods of Estimating the Effects of Melanophore Changes on Animal Coloration," *Biological Bulletin* 84.3 (June 1943), pp. 273–84.

23. Len Lye, "The Man Who Was Colorblind: An Example of Chromatic

Continuities," *Sight and Sound* 9.33 (Spring 1940), p. 6.

24. Poulton, "The Place of Mimicry in a Scheme of Defensive Coloration," p. 305.

25. Len Lye, "Experiment in Colour," in Wystan Curnow and Roger Horrocks (eds.), *Figures of Motion: Len Lye, Selected Writings* (Auckland: Auckland University Press, 1984), p. 47, first published in *World Film News*, December 1936.

26. Katie Zezima, "When Soldiers Go to War, Flat Daddies Hold Their Place at Home," *New York Times*, September 30, 2006, p. A8.

27. As one woman described her family's experience interacting with the photographically dressed cardboard cutout: "Much of the time we simply keep moving forward as if there's no hole in our family. It's sheer pretense, as flimsy as a tissue, and I'm not sure how long it's sustainable — or if it will get us through the long days ahead. But it's better than pretending a smiling cutout loves us back." Here the flat daddy comes to represent an outline of what's not there — a stencil that opens only onto "the hole." Alison Buchholtz, "A Father on Poster Board Just Won't Do," *New York Times* April 8, 2007.

28. Poulton, "The Place of Mimicry in a Scheme of Defensive Coloration," p. 305.

29. H. G. Wells, "The Unveiling of the Stranger," in *The Invisible Man* (1897; New York: Dover, 1966), p. 27.

30. Mark Twain, *Following the Equator: A Journey Around the World* (Hartford, CT: The American Publishing Company, 1897), p. 645.

31. Mark Twain, *What is Man?* (New York: De Vinne Press, 1906), p. 57.

32. Siegfried Kracauer wrote in his *Theory of Film* that films sometimes are able to "cling to the surface of things." Siegfried Kracauer, *Theory of Film: The Redemption of Physical Reality* (1960: Princeton: Princeton University Press, 1997), p. 285.

33. War Department of the U.S. Army, *Notes on Camouflage: Edited at the Army War College, September 1917* (Washington, D.C.: Government Printing Office, 1917), p. 17.

34. Adrian Cornwell-Clyne, "Notes on Concealment," 1915, twelve-page manuscript, A. Cornwell-Clyne Collection, Imperial War Museum Department of Documents, IWM DOC 86/53/1, p. 2. Cornwell-Clyne would later become an expert in color and three-dimensional cinematography, publishing classic textbooks on both subjects. During the 1930s, he was spokesman and agent for Dufaycolor film stock and wrote letters to the editor of *Nature* on the subject of

camouflage during World War II.

35. Stephen Kennedy and Alice F. Park, "The Army Green Uniform: Technical Report 68-41-CM," Clothing and Organic Materials Laboratory: United States Army, March 1968.

36. Infrared (IR)-brightened dyes were incorporated into the fabric with the goal of reducing the differential between environmental radiation and that by the soldier's body. Alvin Ramsley and William Bushnell, "Development of the U.S. Woodland Battle Dress Uniform," U.S. Army Natick Research and Development Laboratories, Clothing, Equipment and Materials Engineering Laboratory, January 1, 1981, ADA 096884; Alvin Ramsley and Walter Yeomans, "Psychophysics of Modern Camouflage," U.S. Army Natick Research and Development Laboratories, June 18, 1982, ADA A117491.

37. However, the Department of Defense aims for perfect filmic concealment beyond that of the predator; notably, even the predator's suiting eventually reveals itself to both the human characters inside and human viewers outside the space of the film.

38. Masahiko Inami, Naoki Kawakami, and Susumu Tachi, "Optical Camouflage Using Retro-Reflective Projection Technology," *Proceedings of the Second IEEE and ACM International Symposium on Mixed and Augmented Reality* (2003), p. 1, available at http://www.astrosurf.com/luxorion/Physique/camouflage-taki.pdf. Retranslated. A similar technology has been presented as "Project Chameleo." See Richard N. Schowengerdt and Lev I. Berger, "Physical Aspects of Electro-Optical Camouflage," Presentation at the American Physical Society Centennial, March 23, 1999, available at http://www.chameleo.net/InnEOCam-PC-Final.ppt.

Index

Zone Books series design by Bruce Mau

Typesetting by Meighan Gale

Image placement and production by Julie Fry

Printed and bound by Thompson-Shore